High Temperature Structure Formation and Surface Diffusion of Silver on Silicon Surfaces

High Temperature Structure Formation and Surface Diffusion of Silver on Silicon Surfaces

Der Fakultät für Physik, Universität Duisburg-Essen
vorgelegte Dissertation
zur Erlangung des Grades
Doktor der Naturwissenschaften
(Dr. rer. nat.)

von

Dirk Wall
geboren am 26. Februar 1982
in Würzburg

Bibliografische Information der Deutschen Nationalbibliothek
Die Deutsche Nationalbibliothek verzeichnet diese Publikation in der Deutschen
Nationalbibliografie; detaillierte bibliografische Daten sind im Internet über
http://dnb.d-nb.de abrufbar.
1. Aufl. - Göttingen: Cuvillier, 2012
Zugl.: Duisburg-Essen, Univ., Diss., 2012

978-3-95404-271-5

© CUVILLIER VERLAG, Göttingen 2012
Nonnenstieg 8, 37075 Göttingen
Telefon: 0551-54724-0
Telefax: 0551-54724-21
www.cuvillier.de

1. Auflage, 2012
Gedruckt auf säurefreiem Papier

978-3-95404-271-5

Gutachter:
PD Dr. F.-J. Meyer zu Heringdorf
Prof. Dr. K. R. Roos
(Bradley-University, Peoria, Ilinois, USA)

Tag der letzten Prüfung: 15. Juni 2012

Hiermit versichere ich, dass ich diese Arbeit selbständig und nur unter Verwendung der angegebenen Hilfsmittel angefertigt und alle wörtlich oder inhaltlich übernommenen Stellen als solche gekennzeichnet habe.
Duisburg, den 26. September 2011

Contents

Part I

General Information

Chapter 1

General Introduction

For a long time thin films have been relevant for technological progress in many fields. In heavy industry, metals are coated by thin films or receive special treatment so that thin films grow for corrosion protection. In optics, thin films are used to custom tune transmissivities and reflectivities of optical devices. Some solid state lasers require systems of layered thin films to achieve the desired optical emission. Without layers of different materials, most semiconductor devices would not be possible and vital physical effects as giant magneto-resistance (GMR) [1, 2, 3] would have never been discovered. However the various applications of thin films also have different definitions to what a thin film is and what surface properties it needs. For corrosion protection and isolation, it is sufficient that the film isolates the metal from the surrounding atmosphere. Here, the surface properties are only relevant if the isolation or corrosion protection are in reactive surroundings. For optics on the other hand, the surface and interfaces are relevant as they effect the reflectivity and transmissivity. However, the films are of sufficient quality if the defects are of smaller order than the order of the wavelength. Especially for surface quality a decent and thorough polishing would be sufficient to achieve the necessary surface roughness and quality for most optical applications. For semiconductor devices, the quality of the films has a much higher importance as single defects can massively influence the electronic properties of devices. For heterolayers in solid state lasers, the homogeneity of the layers has to be at least one order better than those for optical applications.

In physical sciences, many effects have been discovered that change or have properties that change as a function of film thickness. Some examples are thermal conductivity in hetero systems (thermal boundary conductance) of thin films which can depend on the film thickness according to the atomic mismatch model (AMM) [4, 5] or the diffuse mismatch model (DMM) [4, 6, 7, 5] which also depends on the quality of the interface at coverages of a few monolayers. As a second example, the conductivity of various materials can change between insulator, semiconductor and metal as the coverage is changed also in the region of a few monolayers [8, 9, 10, 11, 12, 13, 14]. As a more recent example, new material properties and fascinating effects can occur as layers reach only single layer thickness. These effects can differ so much from the bulk properties that they are considered a different material as it is the case for Carbon and Single Layer Graphene. Here, even a second layer changes the properties from metal to semiconductor as the interaction of two graphene layers leads to the opening of a band gap, even though it is rather small [14].

However, not only film thickness has a strong influence on properties of structures. Also the lateral dimensions can lead to quantum state effects and thereby manipulate the properties of structures as they are used for plasmonic devices [15, 16, 17, 18, 19, 20]

Various approaches and techniques have been taken in the past to investigate film structure and properties. Here, we use a new technique to study surface diffusion and combine it with SEM and SPA-LEED. Our technique is based on either photoemission or reconstruction based contrast in real-time surface sensitive microscopy using either a Low Energy Electron Microscope (LEEM) or a Photoemission Electron Microscope (PEEM).

Part I deals with the basics. First, general basics are presented in Ch. 1. In Ch. 2, the experimental setups are presented. Chapter 3 deals with the special characteristics of Si and Si surfaces.

Part II contains results on Ag growth especially the Island formation on Si(1 1 1).

In Part III, any results involving diffusion and diffusion anisotropy are presented.

Part IV contains the conclusions and outlook as well as the appendix.

1.1 Nucleation and Growth

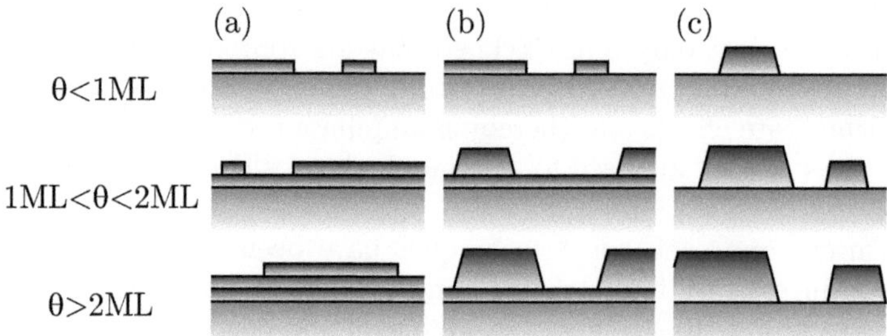

Figure 1.1: Schematic representation of the three crystal growth modes: (a) layer by layer or Frank-van der Merwe, (b) layer plus island, Stranski-Krastanov, (c) island or Vollmer-Weber mode. θ represents the coverage in monolayers (ML)

The considered thin film deposition situation is a special case of epitaxial growth. The term *epitaxy* is used for the growth of a crystalline layer upon (*epi*) a crystalline substrate, while the crystalline composition and orientation of the substrate impose an order (*taxis*) on the orientation of the deposited layer [21, 22]. An epitaxial system has therefore two parallel contact planes of the two crystal structures and parallel crystallographic directions within these planes. The direction the epitaxial layer forms is selected by the constraint to minimize the interface energy which results from the mismatch between the two crystal structures. If a film material different from the substrate material is used, and this is the only case considered in this work, the term hetero epitaxy is used. In general, three different modes of crystal growth are widely accepted for growth near the thermodynamic equilibrium. The growth modes are illustrated in Fig. 1.1 according to Bauer [23].

Figure 1.1 (a) shows the layer by layer or Frank-van der Merwe growth mode, where a new layer is only nucleated after all previous layers are completed. This growth mode occurs when the deposited atoms or molecules are more strongly bound to the substrate than to each other. Figure 1.1 (c) shows the island or Volmer-Weber growth mode where islands are formed and layers are not completed. Islands can, however, grow in both vertical and lateral directions. Figure 1.1 (b) shows the layer plus island or Stranski-Krastanov growth mode where one or more layers are completed before the film roughness increases and islands are formed. This is thus an intermediate case of the other two growth modes. The growth modes can be distinguished according to the balance between the surface free energy γ_D of the deposit, the surface free energy γ_S of the substrate, and the interface free energy γ_{int}. If

$$\gamma_S > \gamma_D + \gamma_{int}, \tag{1.1}$$

the energy balance requires that the area the adsorbate uses is maximal and the atomic layer will grow smoothly in the layer by layer growth mode. If on the other hand

$$\gamma_S < \gamma_D + \gamma_{int}, \tag{1.2}$$

the energy balance requires a minimum of the adsorbate covered area. Therefore, 3-dimensional islands are formed and grow predominantly in height. However, for most scenarios in heteroepity, in the initial stages of layer by layer growth, the deposited material will be strained and and elastic energy will build up. Therefore, Eq. 1.1 will eventually break down after a certain number of layers are grown. The chemical influence of the substrate is vastly screened by the additional layers and the surface free energies γ_D and γ_S are close to identical in the n-th layer. The elastic influence of the substrate remains and therefore is $\gamma_{int} > 0$. This inevitably leads to the case of Eq. 1.2 and therefore Stranski-Krastanov growth as illustrated in Fig. 1.1 (b). The elastic energy can be decreased by the implementation of various crystal defects as for example dislocations if the formation energy is reached before the layer by layer growth collapses.

The early stages of thin film growth involve various steps between the arrival of particles on the surface and nucleation (see Fig. 1.2) and therefore various exchanges of kinetic, vibrational and potential energy. The kinetic energy of an atom arriving on the surface is usually of the order of $\sim 0.1\,\mathrm{eV}$ if the atom was created by evaporation [22]. As an atom approaches the surface, it is accelerated by the surface potential. In order to stick to the surface the atom has to dissipate a considerable amount of energy to place it below the adsorption energy E_{ad} in the order of $1\,\mathrm{eV}$ so that the atom has not enough energy to escape again. The probability of this to occur is then called the *sticking coefficient* of the atom. Once an atom has been adsorbed, it is called an adatom and the next relevant energy scale is the activation energy of surface diffusion of a few tenths of an eV, which is usually smaller than the adsorption energy. Therefore, in the

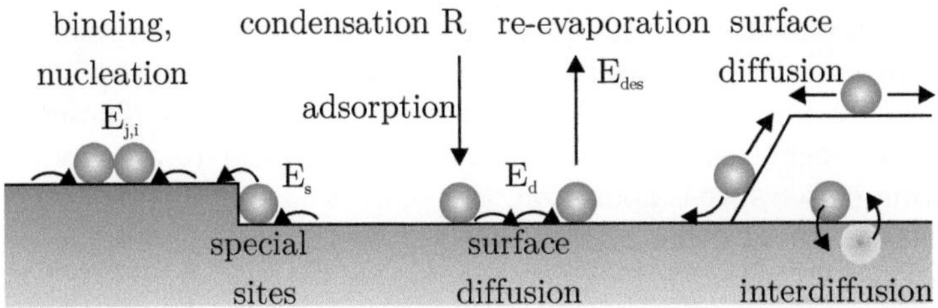

Figure 1.2: Illustration of various possible surface processes occurring during the early stages of thin film growth. This illustration is not exhaustive.

event that a sufficiently efficient transfer of energy from adsorbing atoms to the surface has not taken place, freshly adsorbed adatoms can have enough kinetic energy to move over the surface (see Fig. 1.2) even if the thermal energy of the substrate is not sufficient for diffusion which is then called transient mobility [24]. Of course other atomistic processes are also possible once an atom is adsorbed on a surface. Most of these processes are illustrated in Fig. 1.2.

If we assume a gas pressure p, the molecular weight m, the Boltzmann constant k_B and temperature T, the rate of arrival R is given by

$$R = \frac{p}{\sqrt{2\pi m k_B T}}.$$

Independent of the origin of the arriving atoms, this leads to a density of atoms or molecules on the surface with a number density $n_1(t)$ with a total number of diffusion sites N_0. This leads to the adatom concentration $\theta = n_1\left(t\right)/N_0$ in the residual adatom gas. These adatoms can diffuse freely as described in section 1.2 until they are lost in one of several processes. These processes include nucleation of clusters, attachment to already existing clusters, re-evaporation, attachment to special sites as kinks, steps and other crystal defects, or dissolution into the substrate. For the following description, we will neglect the latter two processes as they are not present at a perfect crystal surface. Each of these processes will be dominated by characteristic times, which if the process is thermally activated are linked by characteristic frequencies and activation energies. The characteristic time for re-evaporation for example is the mean desorption time τ which is given by

$$\tau = \frac{1}{\nu}\exp\left(\frac{E_a}{k_B T_s}\right), \qquad (1.3)$$

where T_s is the substrate temperature and ν is the characteristic surface vibration frequency. Some other characteristic energies are the diffusion energies E_D, binding energy of small clusters of size $j\left(E_j\right)$ until the critical nucleus size $i\left(E_i\right)$ is reached. When larger clusters can decay onto the surface or desorb directly, the associated energies will also become important.

1.2   Diffusion

Diffusion is one of the principal processes of crystal growth as diffusion is necessary for adatoms to find the best incorporation site to form the correct crystallographic structure. As surface diffusion is a thermally activated process, crystal growth morphology depends strongly on the surface temperature.

1.2.1   Classical Surface Diffusion

The most simple view of surface diffusion is the case when adatoms move freely on the surface and the movement can be described as a random walk/Brownian motion $r(t)$ on the lattice of preferential adsorption sites [25]. The mean square displacement grows linearly in time

$$\left\langle (r(t) - r(0))^2 \right\rangle = vl^2 t = 4Dt, \qquad (1.4)$$

where l^2 is the mean square distance covered in a single jump, v is the jump rate and D is the surface diffusion coefficient.

$$D = \frac{1}{4} vl^2 \qquad (1.5)$$

is normalized by a conventional factor of 4 originating in the two dimensional nature of surface diffusion. This normalization leads to coincidence of D and the collective diffusion coefficient D_c in the low coverage limit. We consider single adatom jumps with an initial and a final configuration, where the two configurations are divided by a saddle point which is energetically more unfavorable and thus results in an energy barrier E_D which has to be overcome for adatom movement. The jump rate will then follow the universal Arrhenius law for thermally activated processes

$$v = v_0 \cdot \exp\left(-\frac{E_D}{k_B T}\right). \tag{1.6}$$

The pre-exponential factor v_0 is usually referred to as the attempt frequency and k_B denotes Boltzmann's constant. The attempt frequency resembles the time scale within an adsorption site and $v \ll v_0$ in order for adatoms to move slow enough to resemble the classical diffusion and especially the random walk picture. Therefore, at room temperature (RT) the diffusion barrier should be at least 0.1 eV.

1.2.2   Collective Diffusion

While Eq. 1.4 describes the diffusion of single particles, the collective diffusion of myriads of adatoms can be described by a density profile $n(\mathbf{r}, t)$. The time evolution can then be described by the differential equation

$$\frac{\partial n}{\partial t} = D_c \nabla^2 n. \tag{1.7}$$

The first quantitative description of diffusion on a macroscopic level was given by Fick [26] as a close analogy to the previously by Fourier [27] developed theory of heat flow. Therefore the first Fick's Law correlates the concentration gradient and the heat flux $\vec{J}$

$$\vec{J} = -D_c \nabla n \tag{1.8}$$

for isotropic diffusion. We assume that the movement of each and every particle is independent of the movement of all other particles as long as we imply that the time intervals are not too short [25]. Fick's first law can however be generalized to interacting particles as the concentration gradient can also be understood in terms of the thermodynamic chemical potential μ and the diffusion coefficient relates to a transport coefficient L as follows

$$\vec{J} = -L\nabla\mu. \tag{1.9}$$

This relates mass transport in a linear manner to the gradient of the chemical potential. In this picture the attractive interaction slows down and the repulsive interaction enhances the diffusion rate in comparison to the case without adatom interaction [28]. Equation 1.9 can therefore be rewritten by introducing a chemical diffusion coefficient $D\,(n)$

$$\vec{j} = -D(n)\nabla n, \tag{1.10}$$

where

$$D\,(n) = L\,(n)\,\frac{\partial\mu}{\partial n} = D_j\,(n)\,\frac{\partial\,(\mu/k_B T)}{\partial\ln n}. \tag{1.11}$$

The description is solely dependent on thermodynamic parameters and the jump diffusion coefficient is defined as

$$D_j\,(n) = L\,(n)\,k_B T/n = v\,(n)\cdot l^2\,(n)\,, \tag{1.12}$$

analogous to the diffusion constant of classical surface diffusion (Eq. 1.5). In the limit of low concentration $(n \to 0)$, the term $\partial\,(\mu/k_B T)\,/\partial\ln n$ is equal to 1 and the diffusion constants for classical surface diffusion and diffusion of interacting particles are identical [29].

1.3   Island Formation

As of section 1.1, only R and T_s influence the change of the chemical potential of the surface $\Delta\mu$. Nevertheless, at least the three parameters E_a, E_d, and E_i are needed to describe the early stages of nucleation growth on a perfect surface. However, in practice, perfect surfaces are impossible to prepare. On real surfaces energies of attachment, detachment, diffusion, etc. at special sites E_s may dominate nucleation and growth as these sites may become nucleation sites and change the critical nucleus size [30, 31]. Apart from the so far mentioned processes, there are various other processes possible on surfaces as for example rearrangement, intermixing, shape changes of clusters, and many more. Diffusion thus influences many stages of crystal growth.

An atomistic description of nucleation and growth has been discussed fully in literature [32, 33, 34, 35, 36] and partially repeated in this section. This description is based on the general rate equations [37, 38, 39, 40] and has the form

$$\frac{dn_1}{dt} = R - \frac{n_1}{\tau} - 2U_1 - \sum_{j=2}^{\infty} U_j \qquad (1.13)$$

and

$$\frac{dn_j}{dt} = U_{j-1} - U_j \, (j \geq 2) \qquad (1.14)$$

where n_1 and n_j are surface concentrations per unit area and the U_j are the net rate of attachment of single atoms to j-

sized clusters. By sorting clusters into stable $(j > i)$ and sub-critical $(j \leq i)$ and summing all stable clusters (also referred to as islands)

$$N = \sum_{j=i+1}^{\infty} n_j$$

Eq. 1.13 and Eq. 1.14 can be simplified to yield

$$\frac{dn_1}{dt} = R - \frac{n_1}{\tau} - \frac{d\,(Nw_x)}{dt}, \tag{1.15}$$

$$\frac{dn_j}{dt} = 0\,(2 \leq j \leq i) \quad, \tag{1.16}$$

$$\frac{dn_j}{dt} = U_i - U_c - U_m. \tag{1.17}$$

The last term of Eq. 1.15 is given by the loss of single particles to N stable clusters with an average number of particles w_x per cluster. Equation 1.16 results from the assumption that the growth conditions are such that a detailed balance is established $(U_j = 0)$ for sub-critical clusters. This leads to the Walton relation [41]

$$\frac{n_j}{N_0} = \left(\frac{n_1}{N_0}\right)^j \sum_m C_j\,(m) \exp\left(\frac{E_j\,(m)}{k_B T}\right), \tag{1.18}$$

where C_j are statistical weights resulting from the configuration and contributions to entropy of these j-sized clusters of type m and $n_1 = R{\cdot}\tau$. The last two terms are attempts to deal changing number of clusters due to coalescence by moving (U_m) or simply growing clusters (U_c). The remaining term (U_i) is the nucleation rate of stable clusters which can also be expressed by

$$U_i = J = \sigma_i D n_1 n_i \qquad (1.19)$$

with the surface single atom surface diffusion constant D (see Eq. 1.5) as this can also be written as

$$D = \frac{1}{4}\nu N_0^{-1} \exp\left(-\frac{E_d}{k_B T}\right)$$

and the surface capture number σ, which describes the flux of particles to critical (σ_i) or stable (σ_x) clusters [33, 34, 36, 35]. The capture numbers are rather complicated as they are calculated by Bessel functions [32]. We however assume, that they are slowly varying in the range of $2 - 4$ for σ_i and $\sigma_x \sim 5 - 10$.

At high enough growth temperatures, the detailed balance ($n_1 = R\tau$) is quickly established and therefore $dn_1/dt = 0$ and the last term of Eq 1.15 can be neglected. With the Walton relation (Eq. 1.18) for i, this leads to the high-temperature nucleation rate [32]

$$J = \sigma_i D \left(R\tau\right)^{i+1} C_i N_0^{-(i-1)} \exp\left(\frac{E_i}{k_B T}\right). \qquad (1.20)$$

For lower temperatures, the last term of Eq. 1.15 dominates the growth as re-evaporation is no longer possible. Then, the loss of particles to stable clusters is given by

$$\frac{d\left(N w_x\right)}{dt} = \left(i + 1\right) U_i + \sigma_x D n_1 N + RZ \qquad (1.21)$$

consisting of a term for the nucleation contribution, one term for the term resulting from surface diffusion and the last term yields the amount of atoms lost due to the fact that they directly adsorb at clusters covering the fraction Z on the surface [32]. The nucleation term is usually negligible and the last term is only relevant for large islands or at high temperatures where adsorbed particles desorb before they can be incorporated into clusters. Of course, the shape of clusters will then play an important role as the area fraction will strongly depend on whether the islands grow in 2 or 3 dimensions. This then leads to

$$\frac{dZ}{dt} = \Omega^{2/3} \frac{d\left(N w_x\right)}{dt} \qquad (1.22)$$

or

$$\frac{dZ}{dt} = \Omega \frac{d\left(N w_x\right)}{dt} \left(\frac{\pi N}{Z}\right)^{1/2} \left(1 - \frac{1}{3}\frac{d\left(\ln N\right)}{d\left(\ln Z\right)}\right)^{-1} \qquad (1.23)$$

for 2D and 3D islands respectively. The Ω is the atomic volume of the deposited particles. These dependencies will lead to different dependencies for n_1, $N\left(t\right)$, and $Z\left(t\right)$ for 2D and 3D growth. Nevertheless, the growth coalescence term will then be

$$U_c = 2N \frac{dZ}{dt} \qquad (1.24)$$

for randomly positioned clusters [42]. Using Eq. 1.17, 1.22, 1.23, 1.24, it is possible to calculate maximum stable cluster size as a function of material parameters caused by coalescence from the cluster density [43, 44]. Herein, Z is the only independent variable, where $N\left(Z\right)$ and $t\left(Z\right)$ are the solutions. However, different regimes of nucleation have to be treated separately.

During initial deposition, the transient regime, the population n_1 builds up as $n_1 = R \cdot t$ and all other terms in Eq. 1.15 are negligible. At high temperatures, the regime lasts for $t < \tau$ and $N(\tau)$ is also negligible. For low temperatures, this regime is limited by the incorporation into stable clusters and lasts until $t = \tau_c = (\sigma_x DN)^{-1}$ and n_x depends on the rate R and temperature [45]:

$$\frac{N}{N_0} \sim \left(\frac{R}{N_0 \nu} \right)^p \exp \left(\frac{E}{k_B T} \right),$$

with p and E given in Table 1.1 for various nucleation regimes. Complete nucleation therein means that desorption is neglected while it is significant for incomplete nucleation. For complete nucleation, the growth is thus dominated by diffusive attachment to clusters. For extreme incomplete nucleation, growth will be caused by direct incidence of adatoms onto clusters. The initially incomplete regime is an intermediate regime of the other two, where desorption cannot be neglected, but growth mainly occurs due to the attachment of diffusing adatoms on the substrate [44, 33]. We thus end up with the central result of nucleation theory

$$\frac{N}{N_0} = \left(\frac{R}{N_0 \nu} \right)^{i/(i+2)} \exp \left(\frac{E_i + iE_d}{(i+2) \cdot k_B T} \right) \qquad (1.25)$$

for nucleation of 2D islands with neglected desorption.

Regime		3D islands	2D islands
Extreme incomplete	$p =$	$2i/3$	i
	$E =$	$\frac{2}{3}\left[E_i + (i+1)\,E_a - E_d\right]$	E
Initially incomplete	$p =$	$2i/5$	$i/2$
	$E =$	$\frac{2}{5}\left(E_i + iE_a\right)$	$\frac{1}{2}\left(E_i + E_a\right)$
Complete	$p =$	$i/\left(i+2.5\right)$	$i/\left(i+2\right)$
	$E =$	$\left(E_i + iE_d\right)/\left(i+2.5\right)$	$\left(E_i + iE_d\right)/\left(i+2\right)$

Table 1.1: Parameter dependencies of the maximum cluster density in various regimes of nucleation [32].

1.3.1   Shape of Crystallites

The two dimensional island shape problem is an analogy to the 3D island growth problem and the crystal shapes will reflect the 2D and 3D crystal lattice of the adsorbate. As a first approximation, the variation in island shapes can be related to the hierarchy of the diffusion processes at the island edges. Therefore, thermally activated processes at the step edge may be relevant for the shape formation if they occur on the same time scale as shape formation $\nu_{process} \geq 1\,\mathrm{s}^{-1}$. If a process independent attempt frequency $\nu_0 = 5 \times 10^{12}\,\mathrm{s}^{-1}$ is assumed, the onset temperature of a process with an activation energy E_a can be estimated using the Arrhenius law (Eq. 1.6) to be

$$T_{onset} = 400\,\mathrm{K/eV} \cdot E_a. \qquad (1.26)$$

The hierarchy of kinetic processes on surfaces can be estimated by a simple next neighbor model (nn-model) for the activation energies. The total activation barrier is subdivided into two parts: the energy difference E_{if} between the initial state i and the final state f, and the kinetic barrier E_{kb}. The kinetic barrier is then assumed to be proportional to minimal nn-coordination of the initial and final state and the energy difference is assumed to be proportional to the difference in nn-coordination of the two states. One arrives at

$$E_a = n_f E_{k-nn} + (n_i - n_f)\,E_{nn} \qquad \text{if } n_i \geq n_f \quad (1.27)$$
$$E_a = n_i E_{k-nn} \qquad \text{if } n_i \leq n_f \quad (1.28)$$

where E_{k-nn} is the energy per coordination in the kinetic barrier, E_{nn} is the next neighbor bond strength, and n_i, n_f are the number of initial and final state nearest neighbors, respectively [22]. This simple model then yields an estimate of the onset temperature above which the equilibrium shape can be reached. If we assume a hexagonal crystal and $E_{k-nn} = 0.1\,\mathrm{eV}$ and $E_{nn} = 0.5\,\mathrm{eV}$, this occurs at a temperature of $720\,\mathrm{K}$.

The equilibrium shape of growing crystallites is determined by the minimization of all surface energy terms of the substrate, the island and especially islands facets, which are not accounted for in this simple model. For example different types of steps as they occur on fcc$(1\,1\,1)$ surfaces, where $\{1\,1\,1\}$ and $\{0\,0\,1\}$ microfacets would give the same energy in this simple model [22].

The Equilibrium shape of a crystal is defined as the shape of minimum total surface free energy for a fixed particle number and volume [46]

$$\int_S \gamma\,(\theta, \phi)\,ds = \min! \tag{1.29}$$

with the specific surface free energy $\gamma\,(\theta, \phi)$ for a surface orientation parametrized by the angles θ and ϕ and the entire surface of the crystal S. In analogy to the 3D case, the total step free energy for a 2D island of fixed particle number and area is given by

$$\int_L \delta\,(\phi)\,dl = \min! \tag{1.30}$$

with $\delta\,(\phi)$ being the step free energy per unit length and integrating over the entire edge length L.

The link between energetics and shape is given by the Wulff theorem [47]. For the 2D case, it was proven in [48] and summarized in [49]. The geometric interpretation of the Wulff theorem is the Wulff construction which is reached, if one starts with a polar plot of the step free energy $\delta(\phi)$ (δ-plot) and draws a straight line perpendicular to the radius vector to each point of the polar plot. The inner envelope of the line assembly is then geometrically similar to the equilibrium shape. This geometrical construction is equivalent to the following analytic expression

$$r(\phi) = \min_{\vartheta} \frac{\rho \cdot \delta(\vartheta)}{\cos(\vartheta - \phi)} \tag{1.31}$$

for the shape in 2D polar coordinates (r, ϕ) and the scale factor ρ of the size of the island [49]. Here, the minimization of the angular variable ϑ corresponds to the construction of the inner envelope of the geometrical construction. From Eq. 1.31, we derive

$$\widetilde{\delta}(\phi)\, \kappa_{st}(\phi) = 1/\rho = const. \tag{1.32}$$

with the local curvature of the shape κ_{st} and the stiffness of the step edge [50]

$$\widetilde{\delta}(\phi) = \delta(\phi) + \frac{d^2\delta}{d\phi^2}. \tag{1.33}$$

Equation 1.32 is a local equilibrium condition where the step chemical potential μ_{st} is constant along the island edge. A curved step deviates from its equilibrium of a straight line by the amount

$$\Delta\mu_{st} = \Omega \cdot \widetilde{\delta} \cdot \kappa_{st}$$

in analogy to the Gibbs-Thomson relation for curved droplets [49]. Therefore, the scale factor of Eq. 1.31 is $\rho = \Omega/\Delta_{st}$.

In the case that the equilibrium shape is smooth, the Wulff construction can be inverted and the step free energy orientation dependence can be derived from the shape. From general arguments, the absence of long range order in one dimension may point to the fact that steps form no true facets and that 2D equilibrium shapes are always curved at finite temperatures [51].

Another useful result of this construction (for 3D crystals) is that the distance from the center of the equilibrium shape to the facet r_i is proportional to its surface free energy γ_i per unit area [52]

$$r_i = \text{const} \cdot \gamma_i. \tag{1.34}$$

1.4   Surface Steps, Step Bundles and Facets

One might expect that the shape of a crystal is given by or at least dominated by its lowest energy surface orientation. In reality, this is only one contribution to the shape of a crystal. The total surface free energy of a crystal can be minimized by combinations of higher energy surfaces which reduce the total surface area. Temperature and coverage dependent surface reconstructions can also have an influence on the equilibrium shape of the crystal as they influence the surface free energy of the crystal. The Wulff construction Eqs. 1.31 and 1.34 determine which crystal surfaces are likely. However, in between these surfaces and due to the fact that surfaces are most likely cleaned and prepared at limited temperatures above 0 K, certain deviations from the equilibrium shape will occur as the relaxation of long wavelength modes on crystal surfaces is slow and will hardly reach the equilibrium in a limited amount of time.

Facets are areas of a single (high index) surface orientation where the surface free energy has a local minimum and therefore are usually formed to minimize the surface area. If however, the temperature is raised, the thermal energy can become large enough for the crystal shape to deviate from the equilibrium shape and form soft transitions from one low energy surface to another. Eventually, when the roughening transition temperature T_R is reached or exceeded, the facet will fully disappear and the surface is smoothly rounded [53, 54].The roughening transition is connected to the presence of an indentation on the

γ-plot. In the Wulff construction, the indentation in the γ-plot results in the presence of a facet. The indentation in the γ-plot arises from the existence of a finite free-energy cost (β, per unit length) for the formation of a step. Therefore, the disappearance of a facet implies that the step free energy β vanishes at T_R and a free formation of steps is likely for $T > T_R$. At low temperature, below the roughing temperature of a low index surface, thermal excitation of islands or voids is unlikely and a vicinal surface has the minimum number of steps, given by the macroscopic orientation of the surface. The surface free energy of a vicinal surface can then be estimated by the free energy of the surface steps.

If we consider a simple cubic lattice, in which the surface tension is given by

$$
\begin{aligned}
\gamma(\phi, \theta) \; &= \; \frac{\epsilon_z}{a_x a_y} |\cos\phi| + \frac{\epsilon_x}{a_y a_z} |\sin\phi\cos\theta| + \frac{\epsilon_y}{a_x a_z} |\sin\phi\sin\theta| \\[2mm]
&= \; \frac{\epsilon_z}{a_x a_y} |\cos\phi| \qquad\qquad\qquad\qquad\qquad\qquad (1.35) \\[2mm]
&+ \; \left[\frac{\epsilon_x}{a_y} |\cos\theta| + \frac{\epsilon_y}{a_x} |\sin\theta| \right] \frac{|\sin\phi|}{a_z}, \qquad\quad (1.36)
\end{aligned}
$$

with the energy cost to break a single bond ϵ_i and the lattice constant a_i in each of the orthogonal spatial directions. The vicinality is hereby measured with respect to the nearest facet orientation (θ, ϕ) and the surface free energy is thus also per unit area projected on the nearest facet orientation $f \equiv \gamma/|\cos\phi|$. We then end up with the energy density for vicinal surfaces ($|\tan\phi| \ll 1$) at $0\,\mathrm{K}$

$$f = \frac{\epsilon_z}{a_x a_y} + \left[\frac{\epsilon_x}{a_y} |\cos\theta| + \frac{\epsilon_y}{a_x} |\sin\theta| \right] \frac{|\tan\phi|}{a_z}$$

$$= \gamma_0 + \beta_0(\theta) \frac{|\tan\phi|}{h}, \tag{1.37}$$

where the first term (γ_0) is the energy density of the flat surface, and the last term (β_0) is the energy cost of a step per unit length with height $h = a_z$. At finite temperatures, steps become wavy [55] and form moving kinks along the step edge. Therefore, the configurational entropy of a step S_0 is used to calculate the step free energy

$$\beta = \beta_0 - T S_0 \tag{1.38}$$

If we consider a step along a high symmetry direction and therefore $\theta=0$, the simplest excitation of a kink is ϵ_y (Eqs. 1.35, 1.37). Then, the energy cost can be found in analogy to the Ising model [56]

$$\beta(T) = \frac{\epsilon_x}{a_y} - \frac{k_B T}{a_y} \ln\left(\coth \frac{\epsilon_y}{2 k_B T} \right). \tag{1.39}$$

The configurational entropy generated by thermal kink formation causes the step free energy of an isolated step to decrease with increasing temperature. However, the configurational entropy on real surfaces is always smaller than that of the hypothetical isolated steps. This is caused by the fact that steps exist with a finite density and the assumption that they do not cross which causes a reduction of entropy when steps get close to each other. Thus, the free energy depends on the step-step distance as well as temperature. This terrace width dependence was first calculated by Gruber and Mullins [57] in terms of a wandering step, which movement is limited by two neighboring fixed walls. As a result, the configurational entropy depends on the separation of two walls $2w$

$$S\left(T, w\right) = S_0\left(T\right) - g'\left(T\right) \cdot w^{-2}, \qquad (1.40)$$

where S_0 is the entropy of the step without walls and g' is a width independent constant. For vicinal surfaces, the separation of the walls is given by the mean step-step distance $h/\left|\tan\phi\right|$ and shows the same behavior [58]. Then, with Eq. 1.38, the projected free energy density f becomes

$$\begin{aligned} f\left(T\right) &= \gamma_0\left(T\right) + \left(\beta_0 - TS_0\left(T\right)\right)\left|\tan\phi\right|/h \\ &+ \left(g'\left(T\right)T/h^3\right)\left|\tan\phi\right|^3 \qquad (1.41) \\ &= \gamma_0\left(T\right) + \beta\left|\tan\phi\right|/h + g\left|\tan\phi\right|^3, \qquad (1.42) \end{aligned}$$

with the free energy cost per unit length β to create an isolated, thermally wandering step of height h and the step interaction parameter $g = g'T/h^3$. The step interaction parameter is explicitly depending on the step diffusivity or its inverse, the step stiffness which is determined by the simplicity at which thermal kinks are excited on the step edge. The stiffness $\tilde{\beta}$ is a measure of a step's tendency to straighten and therefore to reduce the number of kinks. The stiffness is related to the step-free energy according to [59, 60, 61]

$$\tilde{\beta}\left(\theta, T\right) = \beta\left(\theta, T\right) + \frac{\partial^2\beta}{\partial\theta^2}. \qquad (1.43)$$

Entropically wandering steps can therefore be expressed using the stiffness as a parameter for the entropic step interaction term g [58, 61, 62, 63]

$$g\left(\theta, T\right) = \frac{\left(\pi k_B T\right)^2}{6h^3 \tilde{\beta}\left(\theta, T\right)}.$$

(1.44)

On real surfaces, elastic and dipole interactions may also be present giving rise to similar inverse square dependence on terrace width [49]. Therefore, the term g will also include the energetic terms apart from the entropic terms. If there is energetic step-step repulsion proportional to A/w^2, the step interaction parameter g becomes [49, 64, 62, 63, 52]

$$g\left(\theta, T\right) = \frac{\left(\pi k_B T\right)^2}{24h^3 \tilde{\beta}\left(\theta, T\right)} \left[1 + \sqrt{1 + \frac{4A\tilde{\beta}\left(\theta, T\right)}{\left(k_B T\right)^2}}\right].$$

(1.45)

1.4.1   Facetting and Step Bunching

Herring [65] showed explicitly, that only facets that are part of the equilibrium crystal shape are thermodynamically stable. Some angles where the surface tension is sufficiently anisotropic, are not present in the equilibrium crystal shape. If a surface is prepared to represent one of these special angles $\hat{n}$ macroscopically, the surface will equilibrate and break up into facets that are represented as part of the equilibrium crystal shape. These facets will, however, not grow large to only represent the two most favorable facets, but the facets will form on the length scale of atom mobility and thus produce a hill and valley structure. Such a hill and valley structure is usually composed of an alternating sequence of the two most favorable facets, thereby conserving the macroscopic orientation of the crystal surface. The surface free energy γ of the entire surface is thus lowered by faceting, even though the surface area A increases

$$\sum_i \gamma_i A_i \leq \gamma A. \tag{1.46}$$

A full thermodynamic description is given in [66, 67]. The central results are the Helmholtz free energy F^s and the excess grand potential Ω^s

$$F^s\left(T, A, N^s_{i \neq 1}\right) = \gamma A + \sum_{i \neq 1} \mu_i N^s_i,$$

$$dF^s = -S^s dT + \gamma dA + \sum_{i \neq 1} \mu_i dN^s_i,$$

and

$$\Omega^s\left(T, A, \mu_{i \neq 1}\right) = \gamma A,$$

$$d\Omega^s = -S^s dT + \gamma dA - \sum_{i \neq 1} N^s_i d\mu^s_i.$$

The restriction results from the definition of the surface. The surface tension is the Helmholtz free energy per unit area for a one component system

$$\gamma\left(T\right) = \frac{F^s}{A} \tag{1.47}$$

and the grand potential per unit area for multi component systems [52]

$$\gamma\left(T, \mu_{i\neq 1}\right) = \frac{\Omega^s}{A}.$$

(1.48)

Therefore, facetting is allowed, if the excess surface free energy is minimized (see Eq. 1.46). In analogy to phase separation, this thermodynamic argumentation leads to the definition of a reduced surface free energy as in Eqs. 1.35 and 1.37 [52, 68]

$$f = \frac{\gamma\left(\hat{n}_i\right)}{\hat{n}_i \cdot \hat{z}} = \frac{\Omega^s}{A}.$$

(1.49)

If we now describe the surface again using the step density $\rho = \tan\phi$, the reduced surface free energy (Eq. 1.41) can be rewritten as

$$f = f_0 + \beta\rho/h + g\rho^3$$

(1.50)

with a surface free energy of a reference plane f_0 which is equal to the surface tension of the reference plane γ_0. Again, the last term is responsible for the incorporation of entropic repulsion and energetic interaction between neighboring steps.

1.4.2 The Influence of Steps on Surface Diffusion

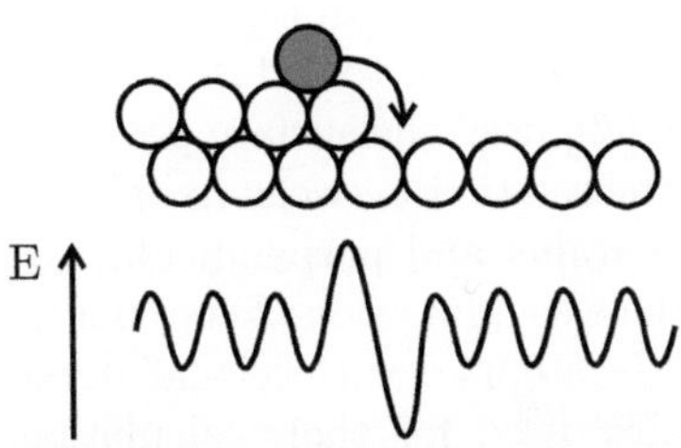

Figure 1.3: The upper part shows the atomistic model of terraces and a step edge in side-view with a descending adatom (shaded). The lower part of the image illustrates the energy landscape for hopping.

The influence of steps on diffusion have some common features for different material and vicinalities. If we resume the Eqs. 1.27 and 1.28 for the activation energy of atom movement depending on the coordination of initial and final state, some results can be drawn. As the number of next neighbors on a terrace differs from the number of next neighbors at a step edge on the lower or upper terrace, steps will have an influence on diffusing adatoms as the diffusion constant follows the Arrhenius law (Eq. 1.6) for thermally activated processes. Therefore, diffusing adatoms can be trapped at lower edge sites and especially kink sites, as the number of next neighbors is higher with respect to the terrace. On the other hand, diffusion atop a down-step edge is favored, as the number of next neighbors is lowest here. This is illustrated in Fig. 1.3. Adatoms have to overcome the additional barrier and may be bound at the lower side of the step edge where the binding is strongest. Each of these processes has to be overcome thermally in order for an adatom to diffuse over a step edge.

Natori and Godby [69] used the surface potential profile (see Fig 1.3) to calculate the relation between hopping rates and the diffusion coefficients parallel and perpendicular to a step on a regularly stepped surface as a function of step density.Figure 1.4 illustrates the regularly stepped surface and potential profiles that Natori and Godby used for their calculations. The step distance is an integer (n) multiple of the surface lattice constant a.

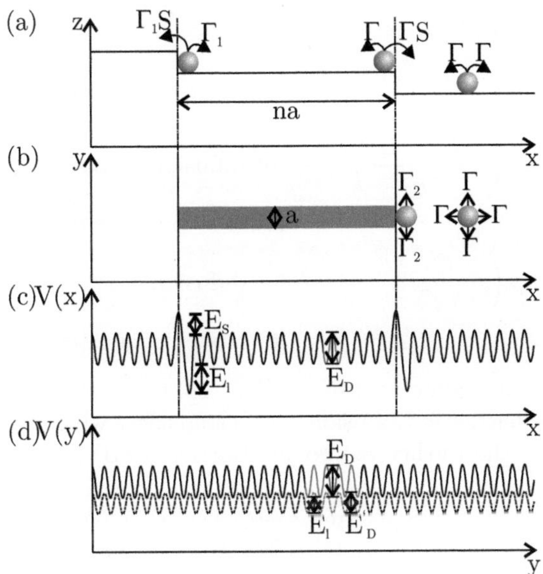

Figure 1.4: A Section (a) and a top-view (b) of a surface with a regular step train and the corresponding potential profiles in x- (c) and y-direction (d) are shown. The dashed line in subfigure (d) illustrates the potential profile along a lower step edge. The step distance is an integer multiple of the lattice constant a (a). The Γ_i's and $\Gamma_i S$ are the hopping rates in the indicated directions [69].

The hopping rates are

$$\Gamma = \nu \exp\left(-\frac{E_d}{k_B T}\right) \qquad \text{for terrace diffusion,} \qquad (1.51)$$

$$\Gamma_1 = \nu \exp\left(-\frac{E_d + E_l}{k_B T}\right) \qquad \text{for diffusion across} \qquad (1.52)$$

$$\text{lower step edge sites,}$$

$$\Gamma_2 = \nu \exp\left(-\frac{E'_D}{k_B T}\right) \qquad \text{for diffusion along} \qquad (1.53)$$

$$\text{lower step edge sites.}$$

S is the so called Schwoebel factor [70, 71] and is related to the extra energy barrier for diffusion across step edges which is often referred to as the Ehrlich-Schwoebel Barrier (ESB) [72, 70, 71]

$$S = \exp\left(-\frac{E_S}{k_B T}\right). \qquad (1.54)$$

Natori and Godby end up with effective diffusion constants for diffusion across D_x and parallel to regular step trains D_y on surfaces [69]:

$$D_x = \frac{n^2}{\left[n - 1 + \frac{1}{S}\right]\left[n - 1 + \frac{\Gamma}{\Gamma_1}\right]}\Gamma a^2,$$

$$D_y = \frac{n - 1 + \frac{\Gamma}{\Gamma_1}\frac{\Gamma_2}{\Gamma}}{n - 1 + \frac{\Gamma}{\Gamma_1}}\Gamma a^2. \qquad (1.55)$$

In this nomenclature, Γa^2 is equivalent to the diffusion constant D_0 for diffusion on a flat surface (see Sec. 1.2).

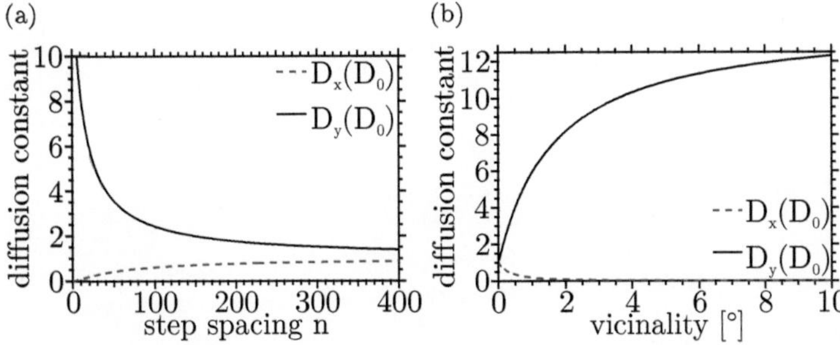

Figure 1.5: Diffusion constants parallel and perpendicular to a regular step train with step distance n as given by Eq. 1.55. For the calculations, the following values were assumed: $E_S = 0.3\,\mathrm{eV}$, $E_l = 0.2\,\mathrm{eV}$, $E_D = 0.6\,\mathrm{eV}$, $E'_D = 0.4\,\mathrm{eV}$, and $T = 900\,\mathrm{K}$. For the conversion of the step distance into vicinality in (b), the step height and lattice constant of $Si(0\,0\,1)$ have been assumed (for values, see tab. 3.2).

Figure 1.5 illustrates the results of the calculations for $E_S = 0.3\,\mathrm{eV}$, $E_l = 0.2\,\mathrm{eV}$, $E_D = 0.6\,\mathrm{eV}$, $E'_D = 0.4\,\mathrm{eV}$, and $T = 900\,\mathrm{K}$ as a function of step distance (a) and as a function of vicinality (b). For the conversion from step distance into vicinality, the surface lattice constant and step height of $Si(0\,0\,1)$ single height steps has been used (see table 3.2).

Chapter 2

Experimental Setup

In this work, various different experimental techniques have been used. Photocmission Electron Microscopy (PEEM) was used to investigate diffusion (an-) isotropy and island growth *in-situ* while Low Energy Electron Microscopy (LEEM) was mainly used to identify and investigate the local reconstructions. The possibility of facetting in the investigated systems was investigated by Low Energy Electron Diffraction (LEED) and Spot Profile Analyzing-LEED (SPA-LEED). For better statistics, some samples were also analyzed *ex-situ* by Scanning Electron Microscopy (SEM). The relevant details of these techniques are described in the following sections. As an overview, the differences of the Ewald constructions are shown in Fig. 2.1.

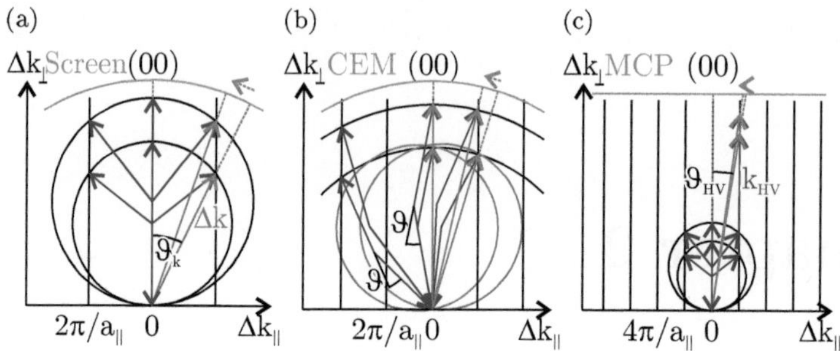

Figure 2.1: Overview of the Ewald-Constructions for LEED in (a) backview LEED systems, (b) SPA-LEED systems, and (c) LEEM instruments.

2.1 SPA-LEED

A SPA-LEED is a high resolution LEED method in which the electrons always have the same angle between incident and diffracted beams of 7° [73]. The diffracted beam is detected by a Channeltron electron multiplier. The pulsed signal from the multiplier tube is preamplified at the UHV-feedthrough, then amplified by an Ortec pico-TIMING Discriminator and later counted by gated counting using a Burr-Brown PCI 20000 instrumentation card in the computer. The scanning with constant angle between incident and diffracted electron beams leads to a modified Ewald-Construction shown in Fig. 2.1 (b). The hereby achieved resolution is far better than the resolution in normal LEED and has a much higher dynamic range due to the Channeltron detector. Due to the higher resolution and dynamic range, far more information can be collected by SPA-LEED compared to normal LEED [74] at a cost of time as this method scans the reciprocal space. The Chamber has been constructed in the course of this work to investigate the Ag induced facetting of vicinal Si(0 0 1) surfaces at the preparation conditions of the PEEM and LEEM measurements on diffusion and diffusion anisotropy.

The sketch of the sample plane in Fig. 2.2 shows the arrangement of the evaporation cells, sample, loadlock system, SPA-LEED and AES setup as it was constructed. The pumping system of the UHV chamber is in a separate plane below the sample plane. The system is equipped with a 6" Turbomolecular pump with a pre-vacuum rotary vane pump attached to the loadlock system. The main chamber is usually pumped only by a combined Titanium sublimation and a $500l$ ion getter pump attached directly to the main chamber.

2.1.1 Normalization of the measured LEED patterns

As the total count rate and the background intensity of LEED patterns depend on the quality and the exact settings of the electron gun, LEED patterns are background corrected by subtracting the mean background intensity of the images and then normalized to the intensity of the mirror $(0\,0)$ diffraction peak. All SPA-LEED images, 1D scans and reciprocal space maps have a logarithmic color scale to make use of and illustrate the extensive dynamic range of the measured data.

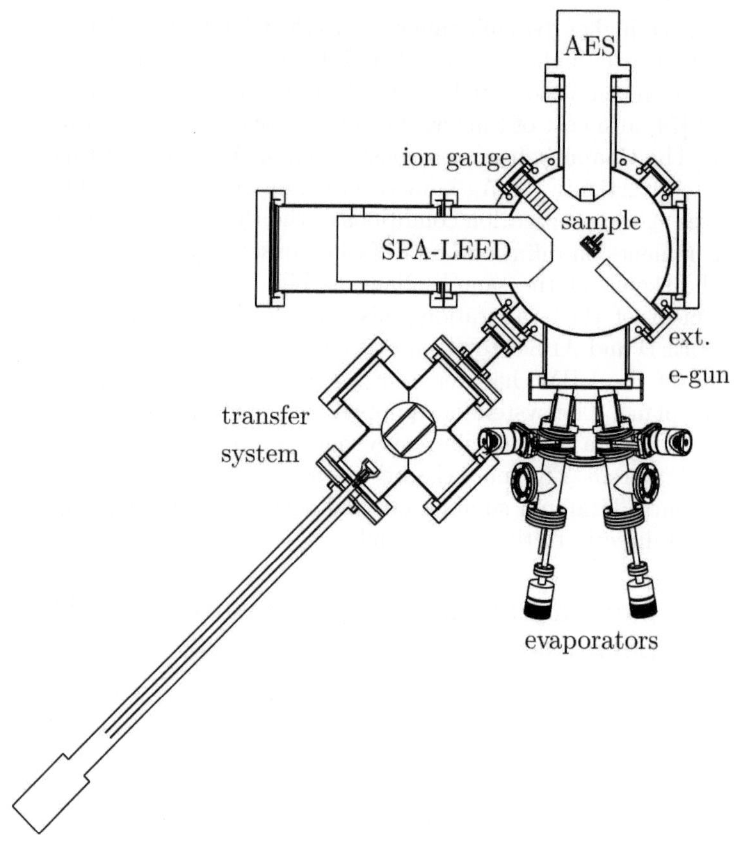

Figure 2.2: Sketch of the sample plane in the new SPA-LEED setup. The external electron gun was not yet installed during the course of this work and the SPA-LEED was not yet modified from cylindrical to conical setup.

2.1.2 Preparation of the LaB_6 emitter

The emission of the LaB_6 emitter of the SPA-LEED changes strongly depending on the contamination of the emitter surface and temperature. Therefore, the emitter was reinitialized on a regular basis by heating the emitter to an emission current of $100\,\mu$A in an O_2 background pressure of $1 \times 10^{-6}\,$bar. The heating current of the filament was therefore adjusted to the emission current until the emission current was stable for $> 10\,$min. By this procedure, the LaB_6 emitter is etched to form a clean and $\mathrm{LaB}_6(1\,0\,0)$ surface with a low workfunction of only $2.69\,$eV [75, 76] to decrease thermal emission to a minimum and achieve the highest possible brightness of the LaB_6 emitter.

2.1.3 LEED Theory

An extensive review of LEED theory, especially the parts that are relevant when using SPA-LEED is given in [74]. This work only uses a small part of the LEED theory for the explanation of the experimental results. The results are sufficiently explained within the kinematic approximation. The relevant parts of the theory [74] are thus presented in the following section.

As for X-ray scattering, the electron diffraction pattern of a crystal with atomic positions at $r\,(n)$ depends on the incident and the final wave vectors k_i and k_f. The wave function can be written using the scattering vector $K = k_i - k_f$ and the structure factor $f\,(n, K, k_i)$

$$\Psi\,(K, k_i) = \sum_n f\,(n, K, k_i) \cdot \exp\left(iKr\,(n)\right). \qquad (2.1)$$

For electron waves, the structure factor combines the surface atoms at $r(n)$ and all underlying atomic columns. The diffraction spot intensity $I(K, k_i)$ is dominated by the scattering cross section and thus the values of the structure factor $f(n, K, k_i)$

$$
\begin{aligned}
I(K, k_i) \quad &= \quad |\Psi(K, k_i)|^2 \qquad\qquad\qquad\qquad (2.2)\\
&= \quad \sum_{n,m} f(n, K, k_i) \cdot f^*(n, K, k_i)\\
&\quad\ \cdot \exp(iK(r(n) - r(m))). \qquad (2.3)
\end{aligned}
$$

As X-rays strongly penetrate matter, the diffraction spots are sharp and for perfect, infinite crystals only δ-peaks as the diffraction pattern reveals the Fourier components of the sample. If the dimensions of the perfect crystal are reduced and the depth of the crystal becomes relevant, the diffraction peaks will smear out in the limited dimension direction. This is always the case for low energy electron diffraction at surfaces, as the penetration depth is only a few monolayers as shown in Fig. 2.3. The electron diffraction pattern will thus be dominated by the smear out of the diffraction peaks in the direction perpendicular to the surface. This smear out of the peaks will usually lead to diffraction rods perpendicular to the surface which are unstructured for the ideal case, where diffraction only occurs at the topmost atoms of a flat surface since the diffraction pattern would thus give the Fourier transform of a δ-function. In the usually observed intermediate case, the diffraction pattern will be modulated by the lattice perpendicular to the surface if the penetration depth is larger, but finite. In addition to this, multiple scattering effects will modulate the intensity along the rods via the structure

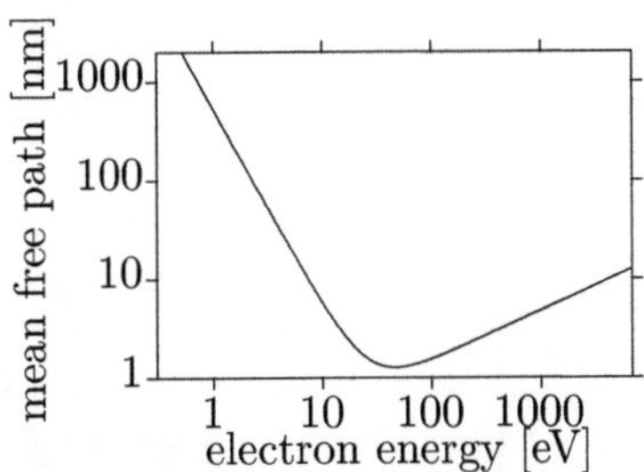

Figure 2.3: The graph shows the mean free path of inelastic electron scattering for elements [77].

factor $f(n, K, k_i)$ depending on incident and final scattering vectors. This is illustrated in Fig. 2.4 (a). The structure factor is different for flat, stepped, and rough surfaces and thus usually the kinematic approximation is used for spot profile analysis.

2.1.3.1 Kinematic Approximation

In the kinematic approximation, all structure factors are replaced by the spatial average of the structure factors

$$f = f(K, k_i) = \langle f(n, K, k_i) \rangle_n . \tag{2.4}$$

This average is independent of special unit cells and the arrangement of neighboring unit cells. This term is thus strictly valid for smooth and flat surfaces, but deviates at positions without translational symmetry. At step edges for example is the arrangement of the surrounding unit cells different. Due to the

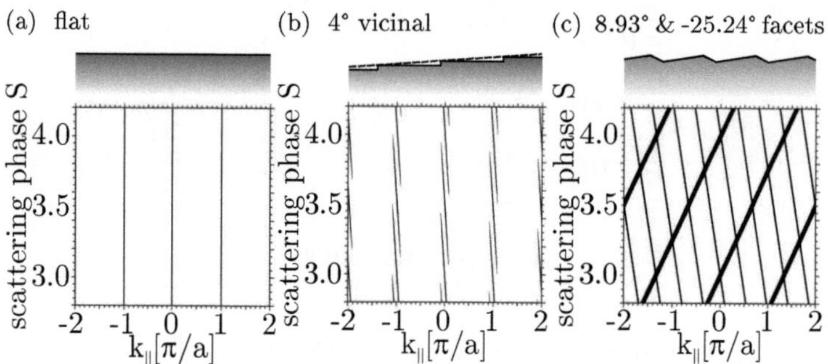

Figure 2.4: Reciprocal space maps (RSM) of characteristic surface morphologies. (a) shows the RSM for a flat surface, (b) shows the RSM of a 4° vicinal surface with a regular step train of double steps, and (c) shows the RSM of a 0° surface which consists of 8.93° and 25.24° faceted areas.[74, 78]

small penetration depth is the deviation limited to a thin band around the step edges and the kinematic approximation is the better, the lower the surface step density is.

The diffraction intensity can be rewritten using Eq. 2.4

$$I\left(K, k_i\right) = F\left(K, k_i\right) G\left(K\right).$$ (2.5)

The dynamical formfactor

$$F\left(K, k_i\right) = \left|f\left(K, k_i\right)\right|^2$$

depending on incident and final wave vectors can thus be separated from the lattice factor

$$G\left(K\right) = \frac{1}{2\pi}\left|\sum_{n} \exp\left(iaK_{\parallel} \cdot n\right) \exp\left(idK_{\perp}h\left(n\right)\right)\right|^{2}$$

which only depends on the scattering vector K. The lattice factor does not influence the intensity modulation of the diffraction spots, it influences the intensity distribution in reciprocal space.

A deviation from a perfectly flat surface will however result in a redistribution of intensity from sharp diffraction spots to spot broadening for rough surfaces, spot splitting for a regular step train as on vicinal surfaces, and new fundamental spots for facets while conserving the total intensity of the diffuse and peaked part of a particular diffraction spot. The measured intensity is then normalized to 1 for N surface unit cells, independent of the vertical part of the scattering vector $K_{\perp}$ and the surface morphology $h(n)$

$$\int_{BZ} dK_{\parallel} G_{ij}(K) = 1$$

of the spot with indices i and j. The lattice factor is obtained from the total integral intensity of a diffraction spot i, j

$$\frac{I_{ij}(K)}{\int dK_{\parallel} I_{ij}(K_{\parallel})} = \frac{F(K,k_i)\,G(K)}{\int_{BZ} dK_{\parallel} F(K,k_i)\,G(K)}$$

$$\cong \frac{F(K,k_i)\,G(K)}{\overline{F}(k_i)\cdot \underbrace{\int_{BZ} dK_{\parallel} G(K)}_{=1}} \qquad (2.6)$$

$$\cong \; G(K)$$

The intensity profile is nevertheless still modulated by the form-factor $F(K,k_i)$ as a function of the electron energy E and the incidence angle φ, and the scattering vector K. In measurements, usually, only the scattering vector is varied and it is assumed that the variation of the formfactor $F(K,k_i)$ is much smaller than the variations of the lattice factor $G(K)$. This approximation is the better, the larger the morphological features are, as this confines the area of the diffuse intensity around a diffraction spot.

We can also express the lattice factor as a function of the surface phase function $\phi(K_{\perp},n) = \exp(idK_{\perp}h(n))$

$$G(K) = \frac{1}{2\pi}\left|\sum_{n} \exp\left(iaK_{\parallel}\cdot n\right)\phi(K_{\perp},n)\right|^2. \qquad (2.7)$$

For a determination of the surface morphology $h(n)$, we need to determine the lattice factor $G(K)$ for different values of $K_{\perp}$.

We introduce the dimensionless scattering phase S to replace the vertical scattering vector and the dependence of the atomic step height d

$$S = \frac{K_\perp d}{2\pi}. \tag{2.8}$$

The scattering phase describes the phase difference of incident and final electrons with the electron wavelength

$$\lambda_{electron} = \left(2\pi^2 \hbar^2 m_e e E\right)^{-1/2}. \tag{2.9}$$

Therefore, electrons interfere constructively for integer values of the scattering phase which are thus also called 'in-phase' conditions. Under these conditions, the lattice factor does not show any influence by surface roughness for energies corresponding to the Bragg conditions. For destructive interference, the so called 'out-of-phase' condition or anti-Bragg condition, the influence of the surface morphology is at its maximum. The scattering phase thus depends on the electron wavelength. For the (00)-spot, this leads to

$$S = 2d\frac{\cos\vartheta}{\lambda_{electron}}. \tag{2.10}$$

For rough surfaces, the lattice factor is given by a sum of δ-functions

$$G_{ideal}(K) = \sum_{i,j} \delta\left(K_\| - \frac{2\pi}{a_0}(i,j)\right)$$

which leads to sharp LEED peaks that are only broadened by instrumental limits and the instrumental resolution. This instrumental broadening is the lower limit for the measured spot width. Any non-regular deviation from the perfect position of the atoms such as point defects or thermal motion will redistribute the intensity from the LEED spots to the background intensity. However, the spot profile is unaffected by thermal motion. The thermal decrease of the spot intensity is described by the Debye Waller factor. Morphological defects, nevertheless, have an influence on the spot profiles as the defects affect the periodic pattern detected by LEED.

The simplest deviation from a perfect surface is a two level system without a random arrangement of the two levels. This would for example be the case for a perfectly oriented surface with single layer height islands. The LEED spots are composed of a sharp central spike given by the long range order of the substrate and a diffuse part which results from the irregular order of the islands. The central spike consists of the constructive interference from the islands and the lower layer. Its intensity depends on the coverage of the islands, which results in a fraction of scatterers on the islands and on the layer $G_{00}\left(K_{\parallel}=0,S\right)=G_{ideal}\left(1-2\theta_1\left(1-\theta_1\right)\left(1-\cos\left(2\pi S\right)\right)\right)$. The intensity is modulated with cosine function of the scattering phase S. As described earlier, the system cannot be distinguished from a perfectly flat surface in the 'in-phase' condition as the scattered electrons will interfere constructively from the two layers as they would also do for a perfectly flat surface. As a consequence of this, the interference from the two layers will for all other phase conditions be partially destructive and completely destructive for a system where half of the surface is covered by the islands.

For increasing vertical roughness, i.e. a multilayer system, the situation becomes more complex. In this situation, the lattice factor is given by the Fourier transform of the projection of all surface scatterers onto the vertical (z) axis with the visible fraction $p_h = \theta_h - \theta_{h+1}$ scatterers in level h

$$
\begin{aligned}
G(K_\parallel = 0, S) &= \left| \sum_n \exp\left(iaK_\parallel n\right) \exp\left(i2\pi S h\left(n\right)\right) \right|^2 \\
&= \left| \sum_n \exp\left(i2\pi S h\left(n\right)\right) \right|^2 \\
&= \left| \sum_h p_h \cos\left(2\pi S h\right) \right|^2 \\
&= \sum_{h,l} p_l p_{h+l} \cos\left(2\pi S h\right) \\
&= \sum_h \frac{1}{2\pi} \int_{-\pi}^{\pi} dS\, G\left(S\right) \cos\left(2\pi S h\right).
\end{aligned}
$$

The last term is the sum over the vertical height correlation which can be obtained by a Fourier transform of the central spike intensity, i.e. the measured $G\left(S\right)$ (Eq. 2.6). However, as the phase information is lost when measuring the LEED patterns, more than one solution of the vertical height distribution is possible.

As discussed in Sec. 1.4, surfaces can exhibit regular step trains and may form facets depending on the angle of the surface orientation and possible adsorbates to minimize the surface free energy. For vicinal surfaces with regular step trains, the diffraction pattern can be described by the multiplication of the

Fourier transform of the super lattice given by the regular step train and the Fourier transform of a single terrace. The Fourier transform of the periodic arrangement of steps is an array of diffraction rods perpendicular to the macroscopic surface orientation. The separation of this array is $\Delta k = 2\pi/$terrace width $= 2\pi/na_0$ and maybe used to determine the lateral dimension of steparrays or facets. The fundamental rods resulting from the single terrace are broadened due to the limited width of the terrace

$$A\left(k\right) = \frac{\sin^2\left(\frac{N}{2}ka_0\right)}{\sin^2\left(\frac{1}{2}ka_0\right)}. \qquad (2.11)$$

This situation is illustrated in Fig. 2.4 (b) showing some exemplary reciprocal space maps (RSM). The inclined rods that are perpendicular to the macroscopic surface orientation intersect the center of the rods perpendicular to the surface in all 'in-phase' conditions as the electrons are then insensitive to the stepped surface. However, a spot splitting may still be observed as the array of steps is still present and influences the formfactor. Nevertheless, if the rods also intersect in other phase conditions, this is due to a different superstructure in vertical direction as it is the case for regular multi-stepped systems.

The description of a faceted surface can basically be understood in the terms just illustrated for vicinal surfaces. Facets can be understood either as surfaces with a characteristic atomic arrangement or as surfaces with an extremely high step density while the atomic arrangement at the steps is usually disturbed by the high step density. To describe a surface using facets, the macroscopic orientation needs to be conserved. The simplest case would thus be a surface with two directions of facets, one leading upward and one leading downward which in total add up to the macroscopic surface orientation. These areas are of limited size as facetting needs a large amount of mass transport for any deviation from the initially flat surface(see Sec. 1.4.1). Figure 2.4 (c) shows a diffraction pattern of a surface with low index facets. Each facet results in a complete set of lattice rods perpendicular to the facet orientation. The diffraction rods intersect in all 'in-phase' or Bragg-conditions. The rods have intensity throughout the diffraction pattern and not only in the vicinity of the fundamental rods as a consequence of an extremely small step-spacing and Eq. 2.11. As the diffraction pattern no longer shows lattice rods perpendicular to the surface, all diffraction spots in a 2D diffraction pattern will move as a function of electron energy unless the RSMs are recorded perpendicular to such a facet direction.

2.2 LEEM/PEEM

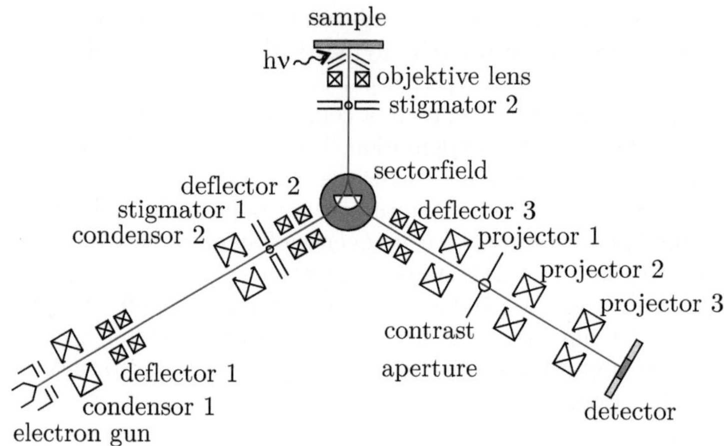

Figure 2.5: Sketch of the course of the electron beam in the IBM LEEM I. The course of the PEEM image originates at the sample and takes the same path through the projection column.[adopted from 79]

Two experimental setups capable of LEEM and PEEM have been used in the course of this work, an IBM LEEM I [80] and an Elmitec PEEM 3 [81] with a 180° energy analyzer which has been upgraded to a LEEM instrument in the course of this work.

Figure 2.5 shows a sketch of the course of the electron beam in the LEEM I instrument. The course of the PEEM 3 instrument has additional electron optical elements on the projection path, nevertheless, the principle of image generation is similar.

Both systems are UHV-instruments and reach base pressures of below 3×10^{-10} mbar. For PEEM, the samples were illuminated by a commercial $150W$ mercury discharge lamp manufactured by LOT-Oriel. For photoemission of Silicon, a photon energy of at least the work function of Si ($\phi \approx 4.8$ eV depending on doping [82]) is necessary. Therefore, the only relevant spectral line supplied by the discharge lamp has a wavelength of $\lambda = 255$ nm giving an energy of $E_{photon} = 4.96$ eV. Therefore, photoemitted electrons have an energy of $E_e = E_{photon} - \phi$. The low photoelectron energy makes this method extremely surface-sensitive. The contrast in PEEM is therefore mainly a mixture of contrast based on density of states and workfunction.

In both microscopes, the electrons are accelerated to HV (20 keV for the Elmitec PEEM and 15 keV for the IBM LEEM instrument) for the ease of electron optics as the aberrations are smaller for higher electron energies. For the IBM LEEM I instrument, the electrons are emitted using a Schottky emitter in a commercial Fei-SEM electron gun and immediately accelerated. During the LEEM upgrade for the Elmitec instrument, a LaB_6 emitter and a $120°$ sector was installed. The electrons then pass various electron-optical elements and are deflected in a $120°$ magnetic sector field before they are decelerated and reach the sample as a parallel beam, where they interact with the surface and are reflected or diffracted. All electrons take the same path from the sample to the detector, regardless of the imaging mode used (LEEM or PEEM). The electrons are then accelerated by the sample potential towards the grounded electron optical el-

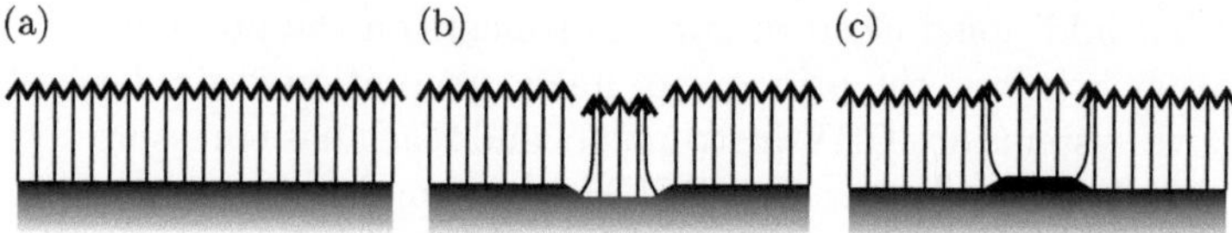

Figure 2.6: Illustration of exiting electrons for different heights and facets. Panel (a) represents a flat surface, panel (b) represents a dent, and panel (c) represents an elevation on a surface. The arrow density relates to the brightness of the image. In LEEM and PEEM, it is possible to find a focusing condition in which for both dent and elevation a dark ring can be imaged arising from its side facets. It is thus hard to distinguish dent and elevation simply by looking at a PEEM/LEEM image.

ements and are deflected by the 120° magnetic sector field and an extra 180° in the energy analyzer for the Elmitec instrument before they are imaged on the microchannelplate detector. The exact settings of the magnetic and electrostatic lenses are used to select an imaging mode and magnification of the instrument. In the electron path, various deflectors and stigmators are used to position the electron beam and minimize image defects.

Since the sample is part of the cathode-lens, the fields in front of the sample are not homogeneous, which is due to the fact that the sample and the sample morphology can be influenced by the acceleration voltage. Therefore, the depth of sharpness is very limited and only very flat samples are suitable for PEEM and LEEM. Even a few nanometers in height can make the difference here as the electrons exit the surface under different angles along facets and the focus is different for different heights (see Fig. 2.6). Electric field enhancement can also play a role if large variations from a flat surface are present.

In LEEM, due to the HV acceleration after the electrons have interacted with the surface, the Ewald construction is slightly modified which is illustrated in Fig. 2.1 (c). In conventional LEED and SPA-LEED, the diffraction spots move radially inwards towards the $(0\,0)$ while in LEEM, practically all of the change in scattering angle is compensated for by the HV acceleration in the objective lens. The original k-space angle θ_k is given by $\sin\theta_k = a^\star/k_{LV}$ with $a^\star = 2\pi/a_\parallel$ and $k_{LV} = 2\pi/\lambda_{electron}$ (see Eq. 2.9) as it is for normal LEED (see Fig. 2.1 (a)) before the electrons are accelerated by the HV. After the acceleration, the k-space angle θ_{HV} is given by

$$\sin\theta_{HV} \quad = \quad \sin\theta_k \frac{k_{LV}}{k_{HV}} = \sin\theta_k \frac{\lambda_{HV}}{\lambda_{LV}} \qquad (2.12)$$

$$= \quad \sin\theta_k \sqrt{\frac{V_o}{V+V_0}} \approx \sin\theta_k \sqrt{\frac{V_0}{V}}, \qquad (2.13)$$

and thus

$$\sin\theta_{HV} = \frac{a^\star}{2\pi}\lambda_{electron}(V) \qquad (2.14)$$

where V_0 is the incident energy of typically $0-100\,\mathrm{eV}$ and V is the microscope potential of typically $15-20\,\mathrm{kV}$. This finally results in the fact, that the measured k-space angle in LEEM is practically constant and the diffraction spots do not move as a function of energy as the Microscope potential is constant.

In the following, the used contrast mechanisms of the instruments are explained.

In the PEEM instrument, the energy analyzer can be used to image only electrons with an energy spread of 0.2 eV by simply inserting a few apertures. As the energy of the electrons can be ramped is it possible to acquire an energy dependent series of images of the surface.

In LEEM, various imaging modes can be used in addition to LEED. One contrast is the so called mirror mode which uses an electron energy, where the electrons are just reflected in front of the surface. The so called mirror energy thereby depends on the local workfunction of the sample and differences in reflectance. This energy is experimentally easy to find since the reflectance decreases dramatically as the electrons reach the sample. This energy therefore gives a good reference and is then used as a reference Energy for LEED and LEEM ($E = 0\,\text{eV}$).

For all other imaging modes of LEEM, it is necessary to have surfaces with regular structures in the length-scale of the electron wavelength, for example crystal surfaces are well suitable.

The bright field mode is a diffraction contrast as only the $(0\,0)$-rod of diffraction is used for imaging. Strictly speaking, mirror mode is also bright field microscopy as only the directly reflected beam is used for imaging, however at null energy, the beam is not diffracted by the sample. To use the bright field mode of the microscope, an aperture is brought into the diffraction plane blanking all electrons that are off the optical axis (and the $(0\,0)$ spot) (see Fig. 2.7 (a)).

For dark-field imaging one would therefore use an inverse aperture which in the case of electrons would not give much infor-

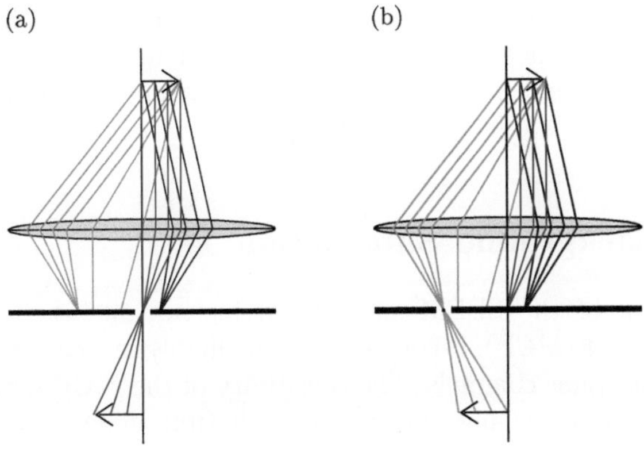

Figure 2.7: Schematics of (a) brightfield imaging and (b) dark-field imaging in LEEM.

mation. However, since the electrons are diffracted, it is possible to use any diffraction spot for imaging (see Fig. 2.7 (b)). In this case, all areas that have the corresponding periodic structure appear bright and all other areas appear dark. If for example a surface consists of only two by 90° rotated periodic areas, the images of the corresponding diffraction spots are inverse of each other. Here, each of the images would only show brightness where the corresponding domains are present. In exactly those regions, the other image would have no intensity as only the rotated domains are present. Nevertheless, since diffracted electrons are used for imaging, the LEEM imaging contrast in brightfield and darkfield imaging modes depends on the phase condition of the electron waves as described in Sec. 2.1.3.1. Therefore, a thought should be spent on what phase condition is best suited for the measurement.

2.2.1   Background Subtraction

The detector of the LEEM and PEEM instruments is a Microchannelplate (MCP). It consists of many highly reactive small electron multiplier channels. The sensitivity of the MCP detectors can therefore change during their lifetime depending on load. Furthermore the amplification changes as a function of applied voltage. Therefore, an absolute number of detected electrons is almost impossible to estimate. The applied load is usually higher in the center of the MCP detector since the $(0\,0)$ diffraction spot and with it maximum intensity has to be on the optical axis for decent real space resolution. The MCP detectors usually exhibit a nonlinear gradient of brightness even when the realspace image is homogeneous. Therefore, background images are recorded with the same MCP settings and similar brightness of the images whenever it is possible to record unpatterned background. This image is then used to estimate the gradient numerically and a difference image is created which is then subtracted from the measured images to correct for the inhomogeneous image amplification of the MCP detector. As it is not always possible to record a background image at the exact same brightness, the difference image is corrected by a possibly nonlinear factor that corrects the gradient the best.

2.3 Sample Preparation

All Silicon samples are polished to $< 0.1°$ of the nominal orientation. Before the samples are inserted into the instruments, they are thoroughly cleaned with acetone to remove carbohydrates and dust and afterwards with propanol to remove acetone and water. The samples are then inserted into UHV and afterwards degassed for several hours at around $600°$ C to remove any leftover contaminations. The sample is then flash annealed to just below the melting temperature of Silicon $(1414°$ C) to desorb the Siliconoxide covering the entire surface. Flash annealing of the sample was repeated a few times to make sure all contaminations have desorbed. Then, the sample heating is lowered and the sample is allowed to cool down again. The Sample is then ready for experiments.

2.4   Temperature Measurement

The sample temperature was measured by a commercial Impac IGA10 infrared pyrometer. Previous work has shown, that the temperature measurement is accurate to $< 10°$ C [83]. In that work, the temperature is measured with a pyrometer for a black body radiator and compared to the temperature measurement using a thermocouple. For Silicon, the pyrometer based temperature measurement of samples heated from the backside using electron bombardment is only reliable above $600°$ C as the infrared absorption of Si increases dramatically below $900°$ C [84]. Above $600°$ C however, the infrared emission of Si dominates transmitted infrared radiation. Even though the transmission does not play a role for resistively heated samples, the emissivity ϵ of the samples is temperature dependent in the spectral range of the pyrometer and is adjusted as a function of the measured temperature using a previously acquired calibration curve [83].

Chapter 3

Silicon and the investigated Silicon Surfaces

Silicon is the second most common element in the earth's crust making up 25.7 % of the crust's weight. It is mostly found as oxides in sand (silica), quartz, rock crystal and in many semi-precious stones [85]. It is most commonly isolated in an electrical furnace where the Si is heated with pure graphite. The formation of unwanted Silicon Carbide can be minimized by keeping the SiO_2 amount high. The Si then reacts with hydrochloric acid (HCl) to form liquid Trichlorosilane

$$Si + 3HCl \rightarrow SiHCl_3 + H_2.$$

The liquid Trichlorosilane is then distilled to form very pure (< 1 impurity per billion molecules) Trichlorosilane. Then, H_2 is added and clean polycrystalline Si forms

$$SiHCl_3 + H_2 \rightarrow Si + 3HCl.$$

Most Si wafers are made from crystals grown by the Czochralski method, where a small single crystal is used as a seed which is dipped into liquid silicon kept slightly above the melting point under protective atmosphere. The seed crystal is then slowly turned and pulled out of the melt and the Si liquid solidifies at the phase boundary. The entire Si rod solidifies in the same orientation as the seed crystal, making it a large single crystal rod. When very clean and well defined wafers are needed, a different, more expensive crystal growth method is used. In this so called Float Zone method, a seed crystal is brought into contact with a clean Si rod in a protective atmosphere. The rod-seed crystal contact zone melts as it is heated inductively. As the chemical potential for impurities is lower in the liquid phase, impurities are carried away from the seed crystal as the entire crystal/rod system is pulled through the inductive heating. The solidification of the molten Si zone is similar to that mentioned above, and the seed crystal has to be rotated to form large single crystal rods.

Wafers are thin slices of these long single crystalline rods and are simply cut from the rods and then polished to the specified orientation, thickness and surface quality. The wafers are then doped by either implanting ions directly by exposure to an ion-beam and then healing crystal defects at higher temperatures (annealing) or by exposing them to doping atom rich gas (for example PH_3 or Diboran (B_2H_6)) at high temperatures $T \approx 1000\,\mathrm{K}$ where the doping atoms dissociate and then diffuse into the wafer, replacing Si atoms. The doping of the wafers used in this work is not relevant for the explanation of the results. If undoped wafers were used for resistive heating, a high voltage would have to be applied to achieve a sufficient current for resistive heating. As Si is a semiconductor, the resistivity drops when the sample is heated and as such, the high voltage would have to be lowered very rapidly to reduce the initial heating once the resistivity starts to drop. We therefore often use doped wafers to eliminate any problems with the resistive heating of many UHV systems as for example the SPA-LEED system. For electron beam heating as it is used in the LEEM and PEEM instruments, this is rather irrelevant. The relevant parameters of the Si crystal are collected in table 3.1.

Property	Value	Ref.
Lattice Structure	Diamond	[86]
Lattice Constant	543.09 pm	[87]
Work Function	4.52 eV	[82]
Band Gap	3.2 eV (direct) 1.12 eV (indirect)	[88]
Resistivity	$1\,\Omega\mathrm{cm} - > 10\,\mathrm{k}\Omega\mathrm{cm}$	[82]
Melting Point	1414° C	[88]

Table 3.1: The table shows the relevant Silicon properties

In this work, $Si(001)$ and $Si(111)$ surfaces as well as surfaces of crystal orientations in between those crystal surface orientations have been investigated. Table 3.2 shows relevant data for the investigated main substrate orientations (001) and (111).

3.1   Miscut Silicon Surfaces

Clean miscut Si surfaces have been used in this work with the nominal orientation of $Si(001)$-0°, 0.2°, 0.8°, 1°, 2°, 4°, 5° misoriented in $[110]$ direction. Furthermore $Si(1111)$, (119), $(11\underline{5})$, (113), (112), (223) and miscut $Si(111)$-0°, 6°, 8° in $[11\overline{2}]$ direction have also been used. All of these substrates can be treated as $Si(001)$ surfaces misoriented in $[110]$ direction as the direction of miscut is the same for all samples in between the (001) and the (111) direction. With increasing amount of miscut beginning at $Si(001)$-0°, the substrate step density increases and steps arrange in a regular manner [55, 92]. At increasing miscut, a transition from single to double steps occurs [93]. For the low index surfaces with increasing vicinality from $Si(001)$ up to $Si(113)$, the number of dimers per terrace reduces gradually [94].

	Si(1 1 1)	Si(0 0 1)
2D Point Group	3	2
2D Bravais Lattice	hexagonal	square
Next Neighbor Distance	$a_0 \cdot \sqrt{\frac{1}{2}} = 384.02\,\text{pm}$	$a_0 \cdot \sqrt{\frac{1}{2}} = 384.02\,\text{pm}$
Atomic Row Distance	$a_0 \cdot \sqrt{\frac{3}{8}} = 332.57\,\text{pm}$	$a_0 \cdot \frac{1}{2} = 271.545\,\text{pm}$
Step Height	$a_0 \cdot \sqrt{\frac{1}{3}} = 313.55\,\text{pm}$	$a_0 \cdot \frac{1}{4} = 135.77\,\text{pm}$
Surface Reconstruction	(7×7)[89]	(2×1)[90, 91]

Table 3.2: The table shows the relevant data for the investigated main substrate orientations.

Part II

Silver and also Gold are widely used to study plasmonics as the plasmon wavelengths of surface plasmon polaritons (SPP) are in the range of visible light. For time resolved studies, Ag/-Si(1 1 1) islands have been used to visualize the movement of plasmon waves [95]. The dispersion relation of light and plasmon waves is different and as a result, an extra impulse vector is needed to excite plasmons on Silver [96]. Therefore, the excitation of SPPs on silver takes place at steps and step edges as the additional impulse vector is available due to the broken symmetry at the edges. Therefore, when SPPs are excited at islands, the waves move beginning at the island edges. For plasmon optics, the shape of such islands is crucial, because if done properly, it is possible to have SPP lenses, beam splitters and other optical elements. However, for the excitation and the output coupling, the shape of the islands and especially the shape of the edges is crucial. The shape can of course be influenced by for example electron beam lithography (EBL), focused ion beam (FIB), or shadow masked deposition. For the understanding of the involved physics however, the system should be as simple as possible. To achieve this, silver is deposited by molecular beam epitaxy (MBE) on Si(1 1 1) samples.

Electromigration continues to be the most serious reliability concern in integrated circuits [97]. During the investigation of electromigration, different mechanisms lead to wire failure. For polycrystalline wires, the wind force dominates and failure occurs at the cathode side of the wires [98, 99, 100]. For single crystalline wires, the direct force dominates and voids occur at the anode side of the wires [101]. To close the link between the polycrystalline and single crystalline wires, attempts are made using bi-crystalline (Ag(0 0 1) and Ag(1 1 1)) islands on Si(1 1 1). These islands are cut into wires and then contacted using FIB. For the fabrication and reproducibility of such wires, the shape, height, crystalline composition and dimensions of the islands have to be controlled.

In Chapter 4, the growth of Ag islands with various rotation angles and crystallinity is analyzed and discussed in context of a coincidence site lattice model.

Chapter 4

Ag Island Formation on Si(111)

4.1 Silver

Silver is one of the materials with the highest electrical and thermal conductance and therefore widely used in microelectronics. Some of the relevant parameters of Ag are collected in Table 4.1. Silver, when deposited on crystalline Si substrates, leads to several coverage dependent reconstructions depending on the substrate orientation. On the Si(1 1 1) surface, the initial (7×7) reconstruction is disturbed and a (3×1) reconstruction starts to form and is fully developed at a coverage of $\frac{1}{3}$ Monolayer (ML) [102, 103]. At higher coverages of up to 1 ML,

Property	Value	Ref.
Lattice Structure	face centered cubic (fcc)	[104]
Lattice Constant	408.53 pm	[105]
Work Function	4.3 eV	[106]
Resistivity	1.471 µOhm cm	[107]
Melting Point	660° C	[108]

Table 4.1: The table shows the relevant Silver properties

a $(\sqrt{3} \times \sqrt{3}) - R30°$ [109, 110] reconstruction forms. At even higher coverages, islands are formed at temperatures above RT. At low temperatures, it is possible to grow Ag films in a layer by layer growth mode [111].

4.2 Island Shapes

Depending on temperature and deposition rate, it is possible to grow islands with various shapes. These different shapes are however still dominated by the crystallographic symmetry of the island. The islands' edges are predominantly oriented with angles of integer multiples of 60° for Ag (1 1 1) islands and 90° for Ag(0 0 1) islands. However, not all islands are single crystalline and islands can have both 60° and 90° angles. This usually means that the island is composed of areas with both crystallographic Ag(1 1 1) and Ag(0 0 1) orientations (see Fig. 4.1). In PEEM, the Ag(0 0 1) and Ag(1 1 1) areas can easily be distinguished as the Ag(0 0 1) areas appear dark and the Ag(1 1 1) areas are bright due to a final state effect [112] which is common for photoemission. In Fig. 4.1, the resulting contrast can be clearly recognized. PEEM images of some possible Ag island shapes are shown in Fig. 4.2. The most common island shape depends on the growth temperature and deposition rate. The temperature dependent shape change illustrated in Fig. 4.2 was acquired while keeping the deposition rate and deposition time constant.

At low temperatures, mainly polygonic islands and bi-crystalline islands form upon deposition. With increasing temperature, first the bi-crystalline islands become unlikely and then, the predominant island shape changes as mainly triangular islands are formed. The data in the graph of Fig. 4.2 does not include the bi-crystalline islands. For better statistics, the data was not acquired with PEEM but was measured ex-situ using the LEO440 SEM. Here, a considerably higher number of islands was in-

	Ag($1\,1\,1$)	Ag($0\,0\,1$)
2D Point Group	3	2
2D Bravais Lattice	hexagonal	square
Next Neighbor Distance	$a_0 \cdot \sqrt{\frac{1}{2}} = 288.87\,\mathrm{pm}$	$a_0 \cdot \sqrt{\frac{1}{2}} = 288.87\,\mathrm{pm}$
Atomic Row Distance	$a_0 \cdot \sqrt{\frac{3}{8}} = 250.17\,\mathrm{pm}$	$a_0 \cdot \frac{1}{2} = 204.27\,\mathrm{pm}$
Step Height	$a_0 \cdot \sqrt{\frac{1}{3}} = 235.86\,\mathrm{pm}$	$a_0 \cdot \sqrt{\frac{1}{2}} = 288.87\,\mathrm{pm}$

Table 4.2: The table shows the relevant data for the main adsorbate crystal orientations.

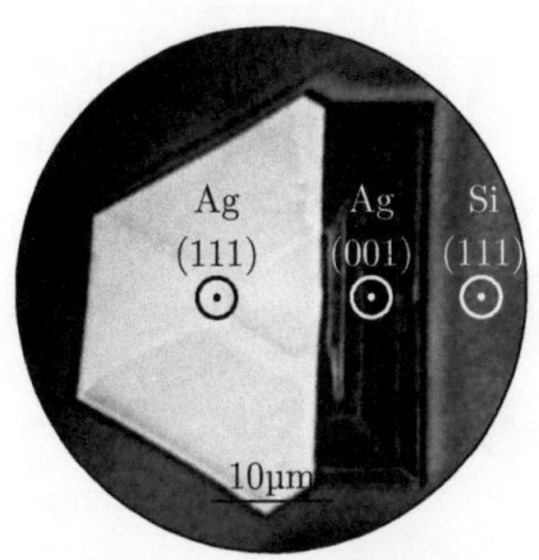

Figure 4.1: PEEM image illustrating the various areas of an island with Ag(1 1 1) and Ag(0 0 1) areas on a Si(1 1 1) surface. The characteristic symmetry of each crystal orientation can clearly be recognized.

spected manually. Even though, the bi-crystalline islands were not included in the data of Fig. 4.2, they have been present and islands with 4-fold symmetrical edges were also present. However, only some of the islands showed the typical contrast in SEM, others appeared homogeneously in the SEM images. Therefore, they were deliberately prepared in PEEM to inspect the difference in the contrast mechanism of the bi-crystalline islands.

It was found, that the crystallinity can change during island growth. Fig. 4.3 shows a PEEM image sequence starting after several minutes of island growth. The different areas are labeled

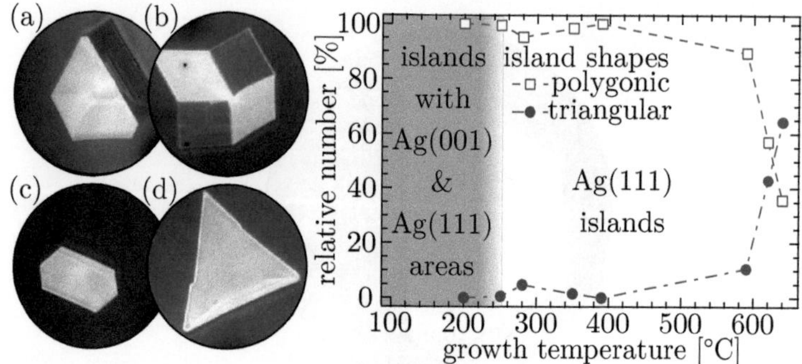

Figure 4.2: Some of the possible Ag island shapes on Si(1 1 1) are shown in (a)-(d) with a 50 µm field of view while the graph illustrates how the predominant island shape changes with temperature. The numbers were counted in SEM images acquired ex-situ for better statistics. The islands were grown using the same evaporator settings for all samples. Islands consisting of both Ag(0 0 1) and Ag(1 1 1) areas have been neglected for the data in the graph as they predominantly form at low temperatures.

in panel (a) where the four fold symmetry of the Ag(0 0 1) area is perfectly visible. During further growth, the contrast of some of the Ag(0 0 1) areas changes and seems to be the same as the Ag(1 1 1) where the contrast change started. This bright area expands rapidly over the Ag(0 0 1) area, until it reaches the opposing edge of the Ag(0 0 1) area (marked with A in panel (i)). Then, the expansion of the bright area slows down considerably

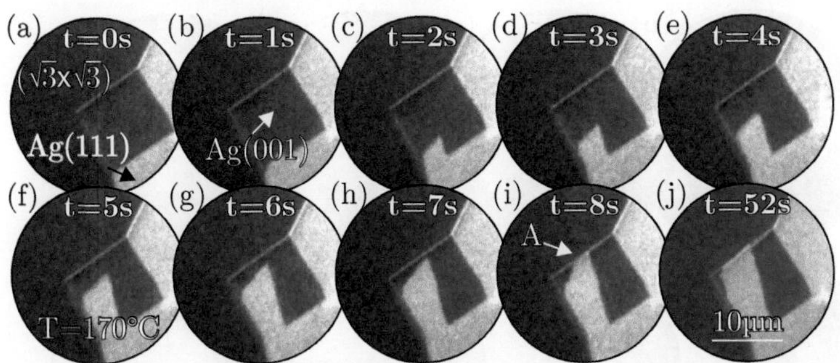

Figure 4.3: Series of images showing the recrystallization of an island consisting of both Ag(0 0 1) and Ag(1 1 1) areas during growth.

as only a small part has changed in contrast until the PEEM image in panel (j) has been recorded much later.

For further investigations of this transformation, bi-crystalline islands were grown deliberately at low temperatures ($< 200°\,C$) while PEEM images were recorded and then heated to above $800°\,C$ as this has also been observed to trigger the transformation. Some of these bi-crystalline islands were analyzed with PEEM and µ-diffraction before and after the transformation to find the cause of this change in contrast. This procedure is more reliable to trigger the transformation than triggering the transformation by simple growth. Figure 4.4 shows the same contrast change of Fig. 4.3. However, this time is the contrast change triggered by a rapid increase in temperature to $T =$

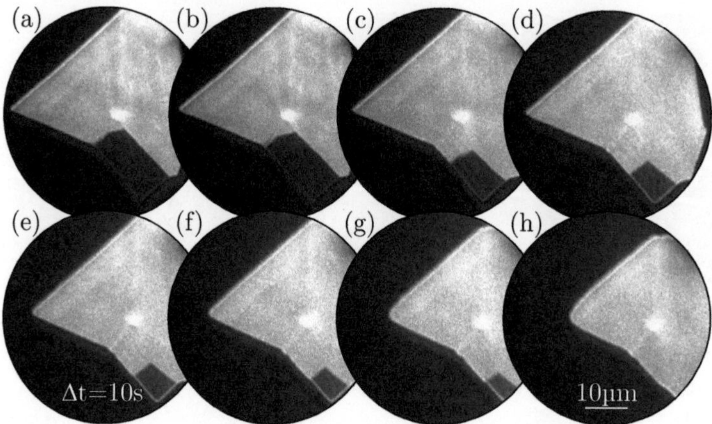

Figure 4.4: PEEM image sequence recorded at a temperature of $T = 775°$ C showing the same contrast change that has already been observed during growth (Fig. 4.3), during the thermal annealing of a bi-crystalline island.

$775°$ C. Panels (a)-(d) show clearly, that the conversion occurs from both boundaries between the two crystallographic areas. In panels (c)-(e), a contrast change of the outer perimeter of the Ag(0 0 1) area can be observed, as the boundary becomes brighter. This is attended by a decrease of the contrast change speed. In panels (e)-(h), the sharp edges of the island become increasingly round as the island slowly decays. The decay mechanism of the islands is treated extensively in Ch. 5. In panel (f), the contrast of the island has totally turned to bright and the

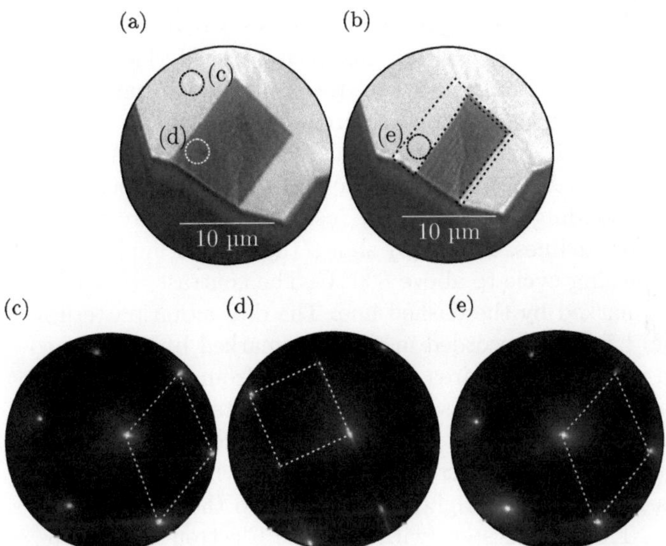

Figure 4.5: Comparison of PEEM images acquired before (a) and after recrystallization (b). The dashed circles mark the areas where the µ-diffraction images (c)-(e) have been acquired. The recrystallized area is marked in (b) by the dashed black line. The recrystallization was triggered by heating the sample to above 800° C for a few seconds. The recrystallized area has the same crystallographic orientation (e) as the original (1 1 1) area (c).

contrast change stops. Figure 4.5 (a) shows a PEEM image of the island after the deposition has been stopped. The dashed circles in panel (a) indicate where the µ-diffraction patterns in panels (c) and (d) have been acquired. The hexagonal symmetry of the Ag(1 1 1) and the four fold symmetry of the Ag(0 0 1) area can clearly be recognized in panel (c) and (d) respectively. The corresponding unit cells of the reciprocal lattice are marked by the dashed lines. Panel (b) shows the same island after a short annealing cycle to above 800° C. The contrast changed in the area marked by the dashed line. The diffraction pattern in panel (e) has been recorded in the area marked by the dashed circle in panel (b). This area now converted from a four fold to a hexagonal symmetry and is aligned to the crystallographic orientation of the original Ag(1 1 1) area. We therefore conclude that the Ag(0 0 1) areas of bi-crystalline islands, where the contrast changes from dark to bright, recrystallize to the Ag(1 1 1) orientation. This is consistent with separate Electron Backscatter Diffraction measurements (EBSD)[113]. In LEEM and PEEM instruments with magnetic lenses, the relative orientation of realspace and reciprocal space images depends on the excitations of each lens. To correlate the PEEM and LEED images, the side facetting of an island was used in µ-diffraction. Then the same lens excitations have been used to record a PEEM image of a bi-crystalline island and the corresponding diffraction pattern.

Figure 4.6 (a) and (b) show a correctly rotated (a) PEEM image and (b) µ-diffraction pattern. The bright area marked by the dashed line in (a) is the electron beam and is thus the area where the µ-diffraction pattern (b) originates. In panel (b), the Brillouin zones of Ag(1 1 1) (dashed, black), Ag(0 0 1) (dotted,

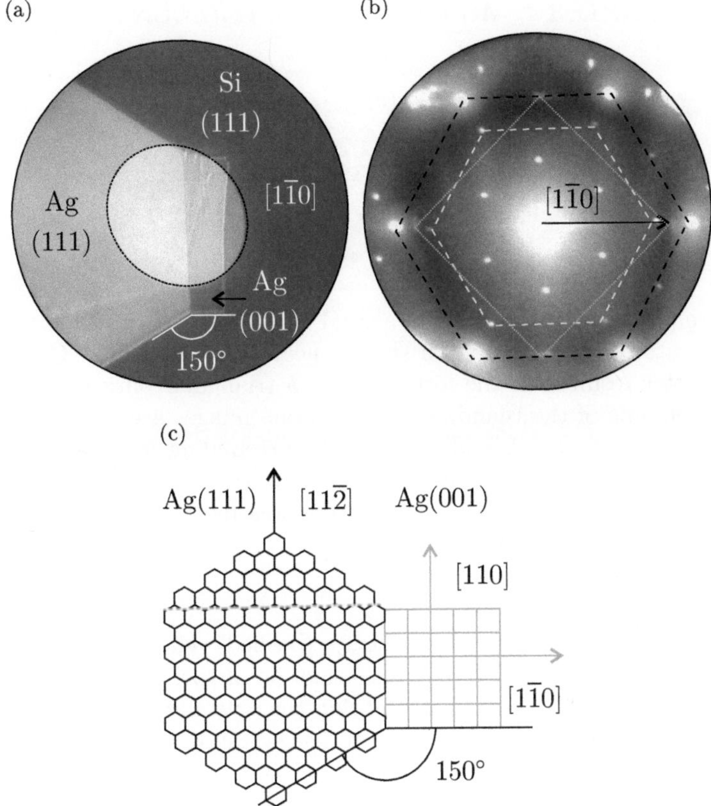

Figure 4.6: (a) shows a PEEM image of a Ag island consisting of both (0 0 1) and (1 1 1) areas. The bright area marked by the dashed line is the area where the µ-diffraction pattern (b) has been recorded. The Brillouin Zones of the Si $\left(\sqrt{3} \times \sqrt{3}\right)$ reconstruction (dashed, yellow), Ag(0 0 1) (dotted, light blue) and Ag(1 1 1) (dashed, black) areas are indicated by the lines in Fig (b). (c) illustrates the crystallographic arrangement within the island as it was extracted from a combination of PEEM and LEED.

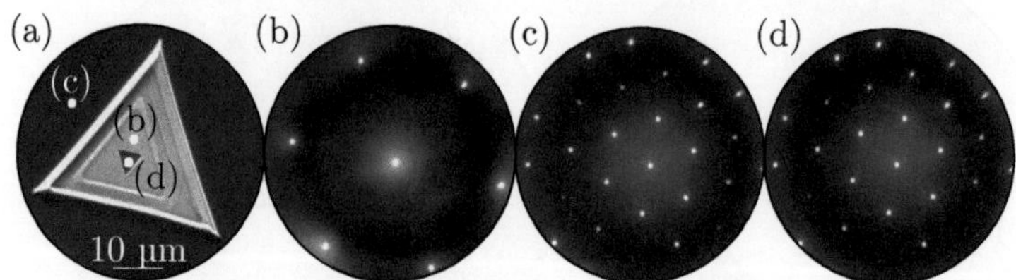

Figure 4.7: (a) PEEM image of a Ag(1 1 1) island grown at low temperatures which was then shortly heated to above 800° C. The latter resulted in the formation on a triangular dark area in the middle of the island. Microdiffraction images are shown, recorded (b) on the island, (c) on the surrounding substrate, and (d) from the hole in the middle of the island as indicated by the bright spots in (a). Each pattern was recorded with an electron energy of 45 eV.

light blue), and the Si(1 1 1) substrate (dashed, yellow) are indicated by lines. The Fourier transformation of these BZs then leads to the Wiegner-Seitz cells and thus the crystalline arrangement illustrated in panel (c). Some of the crystallographic directions are indicated in the images. The grain boundary in between the two areas shows a commensurate ratio of 4:3 boundary sites. Both areas have the $[1\,\bar{1}\,0]$ direction in common, while the lattice constant in this direction relaxes during the recrystallization.

A separate effect that has been found to occur on a regular basis upon a short annealing of the substrate to above 800° C is the growth of dark areas in the middle of Ag islands on Si(1 1 1). Such an island is shown in Fig. 4.7 (a). The bright spots indicate the areas where the µ-diffraction patterns of panels (b)-(d) have been acquired. Panel (b) shows the typical diffraction pattern of a Ag(1 1 1) crystal surface. Panels (c) and (d) show the identical diffraction pattern of the Ag induced Si(1 1 1)-($\sqrt{3} \times \sqrt{3}$) reconstruction. Therefore, the dark area in the middle of the island shows the Ag induced surface reconstruction of the underlying Si(1 1 1) substrate and the only conclusion can be that the island broke up in the middle and a hole formed in the middle of the island that is deep enough to reveal the Si substrate.

4.3   Island Orientation

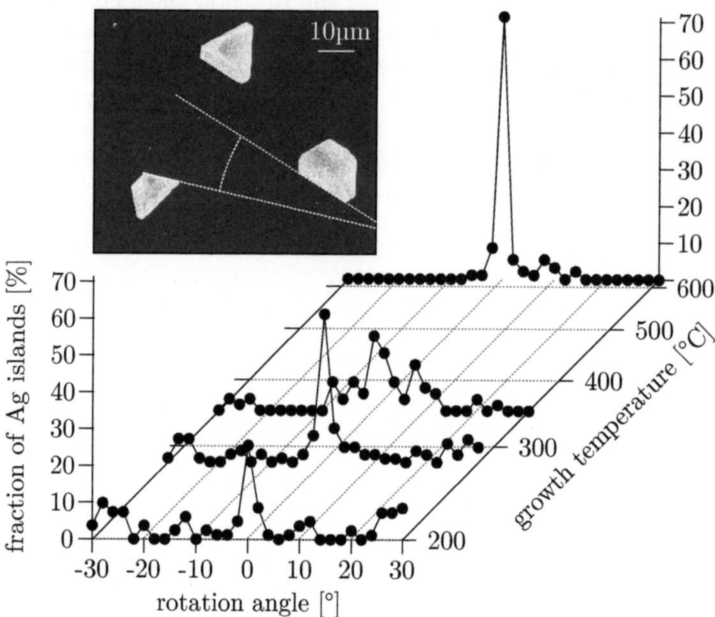

Figure 4.8: The graph shows relative numbers of Ag islands as a function of rotation angle and temperature. The inset indicates how the angles were measured from SEM images.

After deposition of Ag onto the Si(1 1 1) surface, it has been found, that not all islands are rotated in the same direction.

Small and large angle rotations are possible. Figure 4.8 shows the angular distribution of Ag islands measured with SEM as a function of growth temperature. The angles were measured as indicated in the inset. The measurement shows clearly, how the angular distribution of the islands is rather coarse at low growth temperatures and the order increases with increasing growth temperature. At a growth temperature of 600° C, almost all islands grow aligned to each other.

A similar effect can be achieved, when the sample temperature is lowered rapidly during growth. This rapid decrease in temperature can lead to the nucleation of various rotated crystalline areas at the perimeter of preexisting islands. This rather academic situation is shown in Fig. 4.9 (a). The dashed line represents the perimeter of the electron beam that was used to acquire the diffraction pattern shown in panel (b). The two yellow lines therein indicate the donut like area from which a rotational scan was extracted the rotational scan (panel (c)) clearly shows that various rotation angles apart from the expected 3- and 6-fold symmetry are possible. The 4-fold symmetry of the $Ag(0\,0\,1)$ areas are not included in this rotational scan as the diffraction spots are located inside the inner yellow ring in panel (b) the lattice constant of these only slightly visible diffraction spots differs. The statistics and resolution of this island is unsuitable for a numerical analysis of the possible rotation angle as the preparation conditions were not well defined.

Nevertheless, this throws up the question, which rotation angles are likely and which rotation angles are practically impossible. We therefore use the numerical approach of a Coincidence

Site Lattice model and use SPA-LEED for better statistics of the LEED patterns (see 4.4) to explain the possible rotation angles by which island lattices are rotated with respect to the substrate lattice .

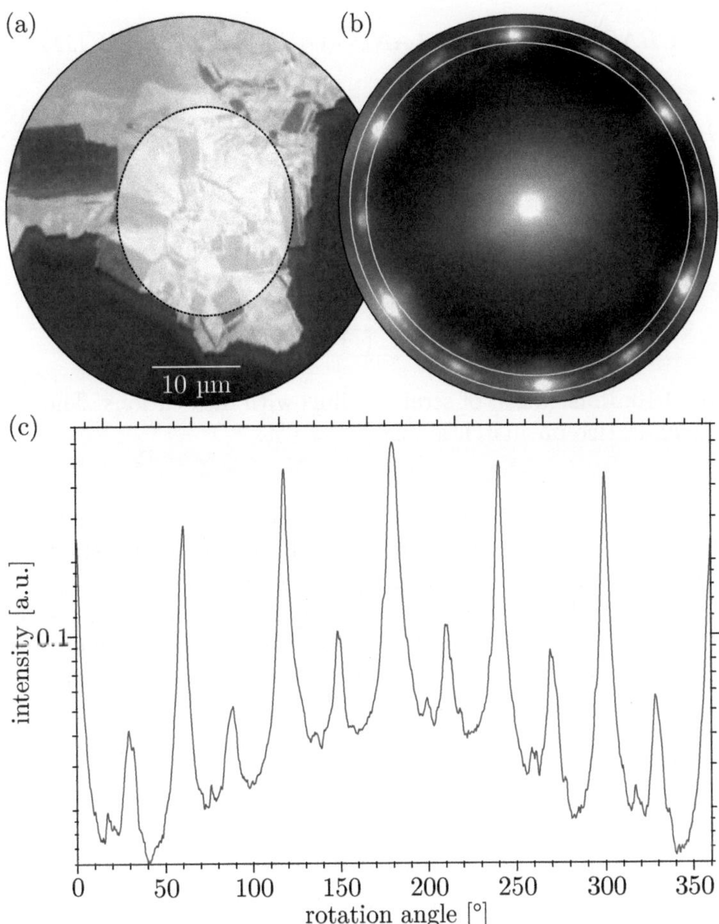

Figure 4.9: (a) PEEM image of a polycrystalline island grown at low temperatures and high deposition rate. (b) µ-LEED pattern acquired from the area indicated by the dashed line in (a). (c) shows the angular distribution of the LEED-intensity in (b) electron beam indicated by the bright area surrounded by a dashed ring.

4.4   Coincidence Site Lattice

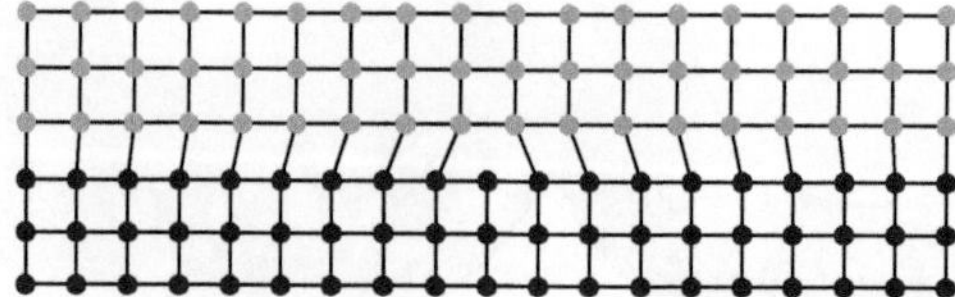

Figure 4.10: Illustration of strained films with dislocations. The displayed lattice mismatch is just below 6%.

Practically all materials differ in lattice constant. Some of them can however still be grown epitaxially. If the lattice constant differs only by a small percentage, the two materials might still grow in a layer by layer growth mode as the first layers of Ge on Si ($\sim 4\%$ lattice mismatch). If the strain becomes too large, dislocations and dislocation networks will form to reduce the strain energy of the film (see Fig. 4.10). However, if the lattice mismatch is large, this would produce many dislocations and is energetically unfavorable. An alternative is an incommensurate superstructure, where the lattices might slightly be altered due to their interaction, however the lattices do not match as they are superimposed. In this situation, there is usually no rational fraction of the lattice constants of the two material systems. An example would be a film which has π times the lattice constant of the substrate (see Fig. 4.11 (a)). Another possibility is that the lattice mismatch can be reduced if one assumes a (possibly rotated) superstructure. This is usually the case if two materials

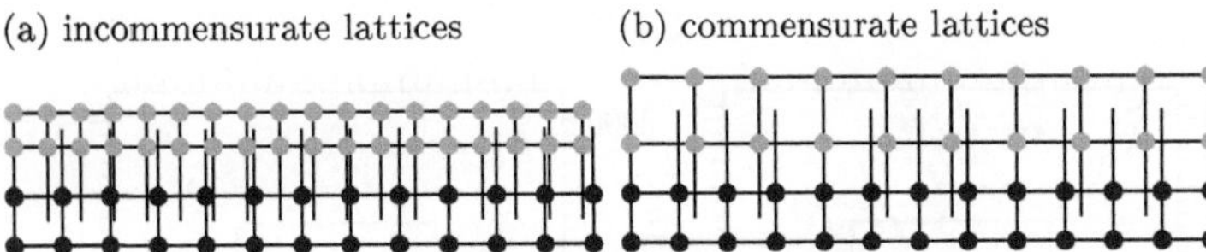

Figure 4.11: Illustration of incommensurate (a) and commensurate (b) lattices. The commensurate lattice has a 4:3 commensurability.

have a lattice constant with small denominator fractions of one another. Therefore, the lattices can for example be superposed and match every 3 lattice constants of the substrate to 4 lattice constants of the adsorbate as it is illustrated in Fig. 4.11 (b). This is described by a commensurate super lattice. If one assumes that two 2D lattices match and form a commensurate super structure, this can be described by a so called coincidence site lattice (CSL) [114, 115, 116]. Analytically the solutions to these CSLs are only possible for homoepitaxial systems as the lattice constants are the same [117, 118]. In this work however, homoepitaxial systems and the possible dislocations are not of interest. Therefore, calculations have been performed to find any possible rotated commensurate structure. Similar calculations have been performed using the coincidence of reciprocal lattice points [119, 120]. Therein, the authors simply counted how many lattice points of one lattice were within a certain radius of the superimposed lattice points. This has also

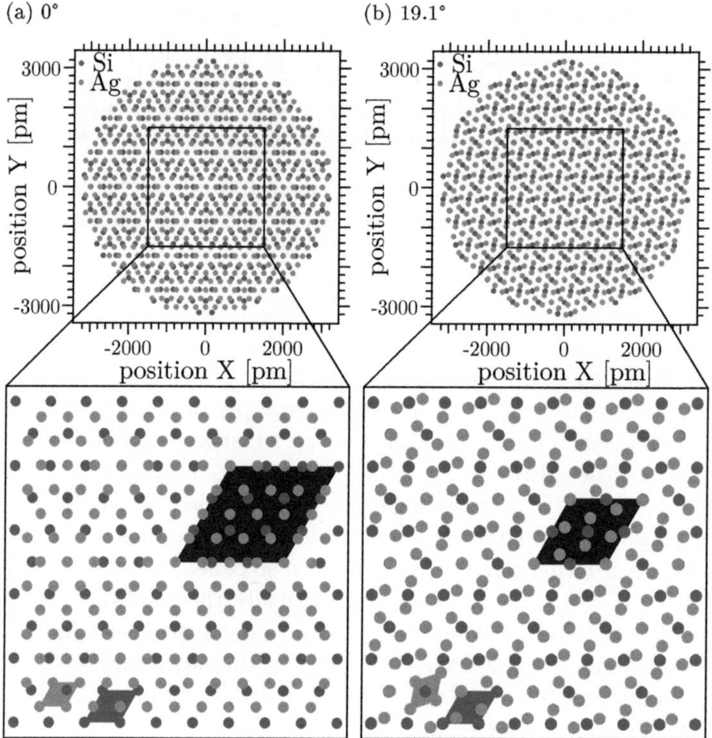

Figure 4.12: Superpositioning of Si(1 1 1) (red) and Ag(1 1 1) (green) lattices with their RT lattice constants and a rotation angle of (a) 0° and (b) 19.1°. The superstructure can clearly be seen and is indicated by the large blue rhombi. The unit cells of the substrate and adsorbate are also indicated by red and green rhombi, respectively.

been tested and gave similar results[1] in this work, however, we used a different approach to generate a parameter that reflects the quality of the overlap of different lattices. This improved the results from the CSL approach. Two point lattices with the symmetry and lattice constants of Ag(1 1 1) and Si(1 1 1) have been generated. Then, for every lattice point (index i) of the Ag lattice, the distances R to the Si lattice points (index j) have been calculated

$$R_{ij} = \sqrt{(x_i - x_j)^2 + (y_i - y_j)^2}. \qquad (4.1)$$

The quality of the overlap is then, due to the coulomb interaction of the two lattices, of the order $\sim \frac{1}{R^2}$ and should be proportional to the inter layer binding energy of the atoms. For this calculation, the lattice has been assumed to be fixed as only the overlap of the lattice constants was evaluated. In real crystals, the lattices would most likely be strained to fit into orientations with low binding energy as long as the strain energy is not larger than the binding energy profit. We end up with the parameter for the quality Q of the overlap of two crystals

$$Q = \sum_{i=0}^{N} \sum_{j=0}^{M} \frac{1}{R_{ij}^2} \qquad (4.2)$$

similar to [121].

As this would give infinite results for $\lim_{R \to \infty} Q$, any distance smaller than $0.1\,\mathrm{pm}$ has been assumed to be equal to

---

[1]The resulting angles were the same, yet was the result more "digital" and the angles of smaller overlap were basically neglected.

(a)   (b)

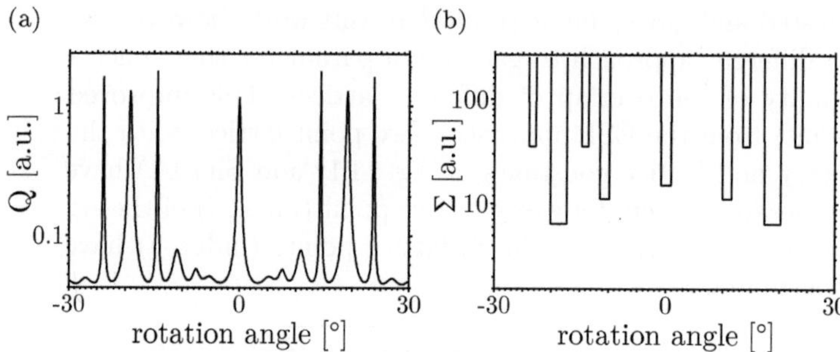

Figure 4.13: (a) Results of the calculation of the quality Q of the overlap of the Ag(1 1 1) and Si(1 1 1) lattices with their lattice constants for a temperature of $T = 160°$ C as a function of the rotation angle between the two hexagonal lattices. High values correspond to good overlap of the lattices. (b) Results of the calculation of the classical CSL parameter Σ of the overlap of the Ag(1 1 1) and Si(1 1 1) lattices with their lattice constants for a temperature of $T = 160°$ C as a function of the rotation angle between the two hexagonal lattices. Low values correspond to good overlap of the lattices.

0.1 pm. As long as the calculation cell is large enough to fit the entire superstructure, the effects depending on cell size should be minimal as the error for missing neighbors for boundary atoms is also of the order $\frac{1}{R^2}$. As the quality of the overlap is estimated as a function of rotation angle between the two lattices, any lattice point outside a cut off radius was eliminated, thus establishing a circular calculation cell. Without loss of generality, it is assumed, that the lattices coincide perfectly in the center of the calculation cell. An example of such rotated superimposed lattices is shown in Fig. 4.12.

The full circular calculation cell is shown in the upper graphs, while the lower plot is a magnification of the area marked in the upper graph. Therein, the CSL super lattice (blue) and the two unit cells of Ag (green) and Si (red) are marked. Panel (a) shows the superposition of non-rotated lattices while panel (b) shows the situation with a relative rotation of $19.1°$.

The superstructure is reminiscent of a moiré pattern, and in principle is a moiré pattern of two superimposed point lattices [122] and could be described by such patterns. The moiré theory however only gives the periodicity of such a superstructure and neither the quality of the overlap nor an idea of the binding energy of superimposed lattices. The size of the moiré could only be used to describe which orientations have small superstructure sizes and would possibly compare to the value

$$\Sigma = \frac{\text{superstructure cell area}}{\text{lattice unit cell area}} \tag{4.3}$$

of the CSL theory which is used as a measure of good overlap [116].

The result of the rotation angle dependent calculation of the quality factor Q is shown in Fig. 4.13 (a). Panel (a) shows the sharp peaks representing good overlap for several rotation angles and was calculated using Eq. 4.2 as a function of rotation angle with the described simulation cell. In Panel (b), the results of the calculations of the classical CSL parameter Σ is shown. The width of the bars shown on panel (b) is directly influenced

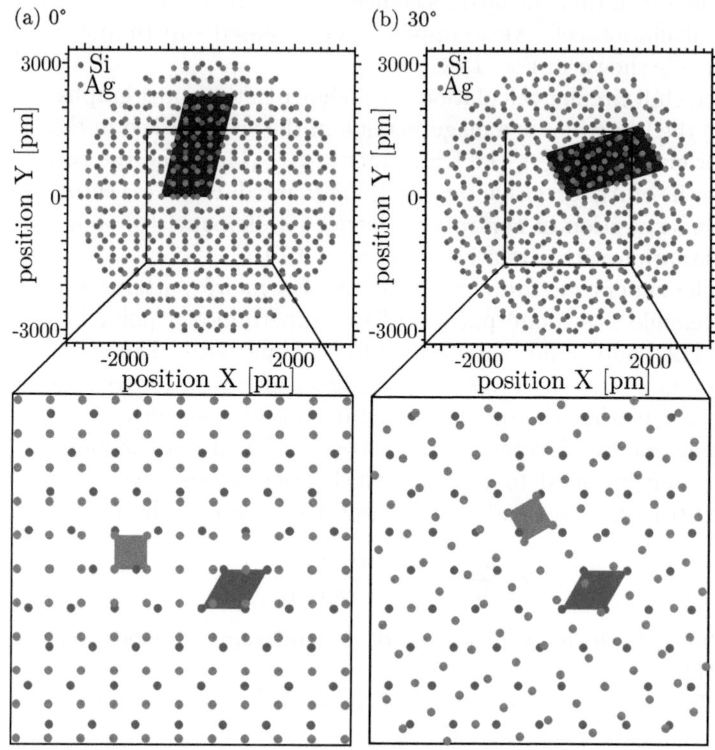

Figure 4.14: Schematics of the CSL for (a) 0° and (b) 30° rotated Ag(0 0 1). The unit cells as well as the CSL supercell are indicated by rhombi. The CSLs were calculated for a temperature of $T = 160°$ C.

by the overlap condition which is used. Here, we assumed that two lattice points coincide, when they are within 10 pm of one another. In this graph, good overlap corresponds to low values of Σ. The simulations agree in terms of where the overlap of the two lattices is best. Figure 4.15 shows the superimposed lattices for which a good overlap has been calculated. For each rotation angle, the CSL lattice unit cells are marked in panels (a)-(f), while the substrate and superimposed lattice unit cells are marked in Fig. 4.12. According to CSL theory, best overlap is achieved by the smallest CSL cell. From the LEEM and μ-diffraction measurements, it is known (see section 4.3) that also $Ag(0\,0\,1)$ crystals may form. Therefore, some of the calculations have also been carried out for the lattices of $Si(1\,1\,1)$ and $Ag(0\,0\,1)$. The most prominent results were $0°$ and $30°$ which is in agreement with the μ-diffraction results. The lattice arrangement of these two lattices is shown in Fig. 4.14, similar to Fig. 4.12. The size of the CSL supercells are in the same range as for $Ag(1\,1\,1)$. Thus, the formation of $Ag(0\,0\,1)$ is predicted to be approximately as likely as the formation of $Ag(1\,1\,1)$ crystals by the modified CSL approach. This result is in good agreement with the μ-diffraction and LEEM results shown previously (see Sec. 4.3)

The next step is to take the rotation angles with good overlap and use them to describe the SPA-LEED measurements. A typical 2D SPA-LEED pattern is shown in Fig. 4.16(a). The 2D scan is similar to the μ-diffraction pattern of Fig. 4.9 only recorded with the larger electron beam spot size on the sample and the higher k-space resolution of the SPA-LEED instrument. A single rotational scan is shown in Fig. 4.16 (b) which was measured separately to achieve a better signal to noise ratio. The

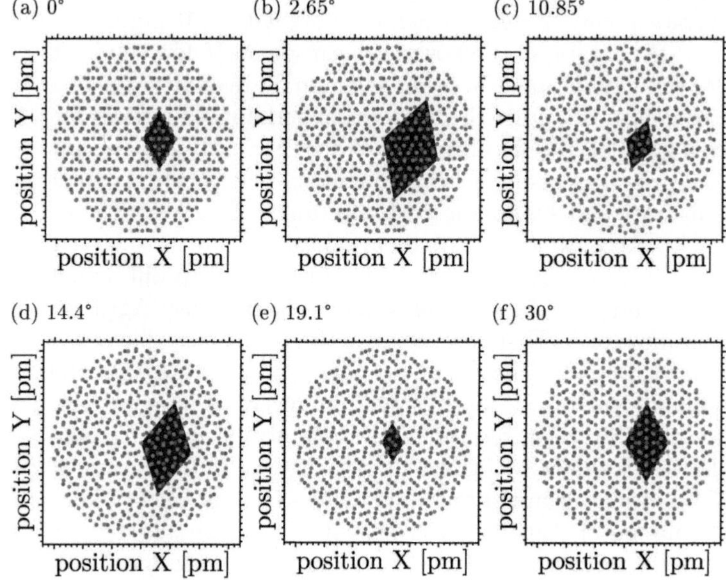

Figure 4.15: Schematics of the CSL for the calculated relative rotation values (a) 0°, (b) 2.65°, (c) 10.85° , (d) 14.4°, (e) 19.1°, (f) 30°. The CSL supercells are indicated by the rhombi. The schematics were created for $T = 160°$ C.

scan shows clearly that several rotation angles are possible. The angles have been marked with the dashed vertical lines. The multi-peak Lorentzian fit (dashed black line) was done using

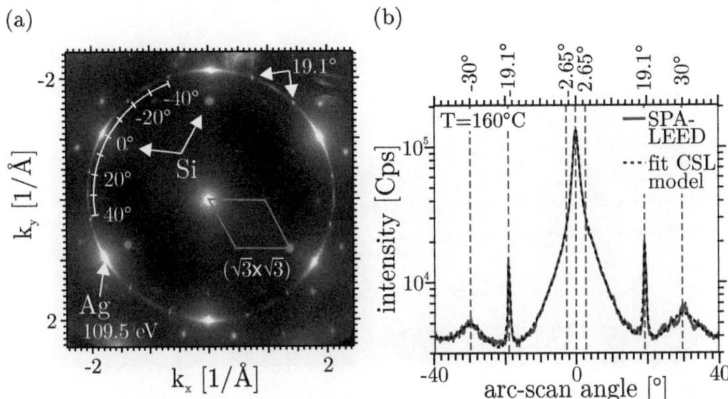

Figure 4.16: SPA-LEED measurement (a) and linescan along the ring, fitted using a multi-peak fit with fixed peak positions from the modified CSL approach (b). The Film was prepared at a growth temperature of $T = 160°$ C before it was allowed to cool to room temperature as quickly as possible, before the SPA-LEED scans were acquired.

the calculated angles for which the CSL structures are shown in Fig. 4.15, only fitting the FWHM and amplitude of each peak while the position was fixed to the calculated values. The angular scale is shown in Fig. 4.16 (a).

Taking this scan, the only result can be, that the model describes the data well, nevertheless, this is not the only thing the CSL model is capable of. It can, if the temperature dependent lattice constants are known, predict the temperature

(a)

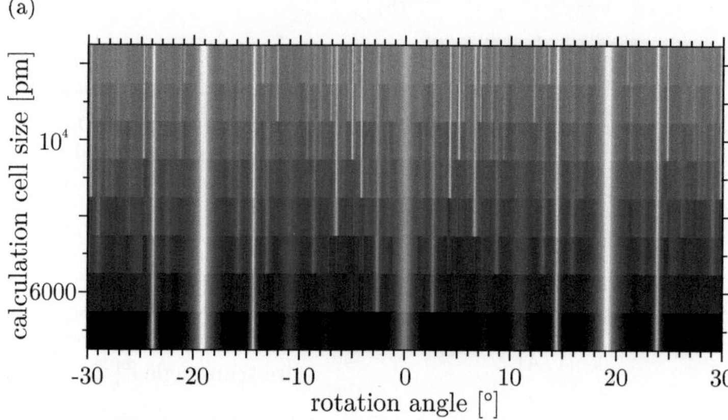

(b)

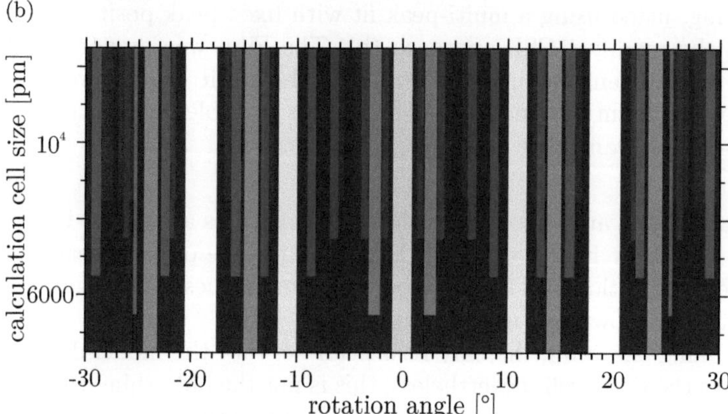

Figure 4.17: Illustration of the size dependence of the calculation cell. (a) shows the quality factor Q and (b) the classical CSL parameter Σ as a function of the rotation angle and calculation cell size at a temperature of $T = 160°$ C. In both images, bright features represent good overlap of the two lattices. The color scale is logarithmic.

dependence of such rotational alignment. However, the model also has some drawbacks and limits, which will be discussed in the following .

For an infinite calculation lattice size, perfect overlap of the two lattices has to occur at some point. Therefore, increasing calculation cell size will lead to lattice sites with perfect overlap and accordingly to extremely high values of the quality factor Q. Thus, the combination of the classical CSL parameter Σ and the Quality factor Q has to be considered. Real lattices can deform to build up strain if this is energetically more favorable as additional coincidence sites can be generated. Within this model, such deformation is neither allowed nor considered. Therefore, this model is more likely to be useful for commensurate structures than for incommensurate structures. Nevertheless, the model is only as good as the knowledge of the lattice constants put into the calculations. Figure 4.17 shows the dependence of the calculations of (a) the quality factor Q and the classical CSL parameter Σ on the calculation cell size and rotation angle. It is clearly visible that new features appear as the calculation cell size increases, while no feature disappears. The small very sharp bright lines in (a) that appear at very large calculation cell sizes of over $10000\,\mathrm{pm}$ are caused by the perfect coincidence that only appears due to the fact that any lattice will coincide at some point. This can be seen in (b) as the corresponding coincidence lines are very dark and as a consequence of the chosen logarithmic color scale, this corresponds to extremely large CSL cells.

For the temperature dependence predictions, it is vital to know the temperature dependence of the lattice 'constants'.

(a)   (b)

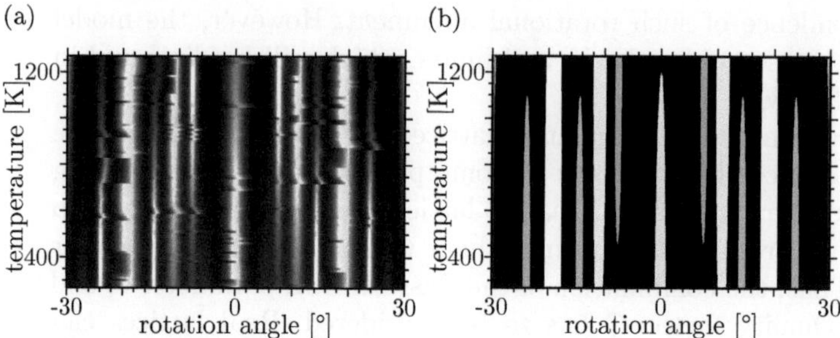

Figure 4.18: Temperature dependent simulation results are shown as a function of rotation angle. (a) shows the quality of the overlap and (b) the classical CSL parameter Σ as a function of temperature and rotation angle of the two lattices with respect to each other. The simulations have been carried out with a discretization of $10\,\mathrm{K}$ and $0.05°$ for rotation. The color-scale is logarithmic and high intensity peaks are cut off to show the fine structure of the results. Bright pixels correspond to a high quality overlap in (a) and a small CSL cell in (b).

Luckily, lattice constants are well documented for almost any material system, including the temperature dependence. All calculations shown up to this point were carried out with the lattice constants for the growth temperature of the SPA-LEED measurement ($T = 160°\,\mathrm{C}$).

For the temperature dependence, we also calculated both the Quality factor Q (Fig. 4.18 (a)) as well as the classical CSL parameter Σ (panel (b)) as a function of temperature (on basis of known thermal expansion of the lattices). Both panels are shown with a logarithmic color scale where bright values correspond to good overlap. Again, both parameters show similar results while the results for the Quality factor Q are less digital and give a better impression of the exact situation. The results show that the coincidence of certain rotation angles varies very slowly with temperature. The angles do not change slowly, but coincidence angles appear and disappear, but not a single value of a "coincidence angle" changes its value. The width and the fading of parameter Σ occurs only due to the fact that we assume overlap if the lattice points are within a previously chosen distance of one another and this changes due to thermal expansion.

To confirm the calculations, SPA-LEED measurements have been carried out for various growth temperatures. The results of this investigation are shown in Fig. 4.19. The scans were taken along the first integer order diffraction ring. The parameters had to be optimized before each measurement because the sample had to be moved in between the measurements for a new sample preparation. Therefore, the recognition of a temperature dependence is very difficult and thus, Fig. 4.19 does not reveal any clear temperature dependence. All scans show a majority intensity at $0°$, $\pm30°$, $\pm60°$ and at some of the other rotation angles from the calculations. However, those are rather difficult do identify. The exact shape of the scans seems to strongly depend on the islands within scanned surface area. Nevertheless, the deposition was adjusted in between the scans for the highest two temperatures and this does have a clear effect on the resulting scans. With decreasing effective deposition rate and increasing diffusion and desorption, the intensity is almost completely limited

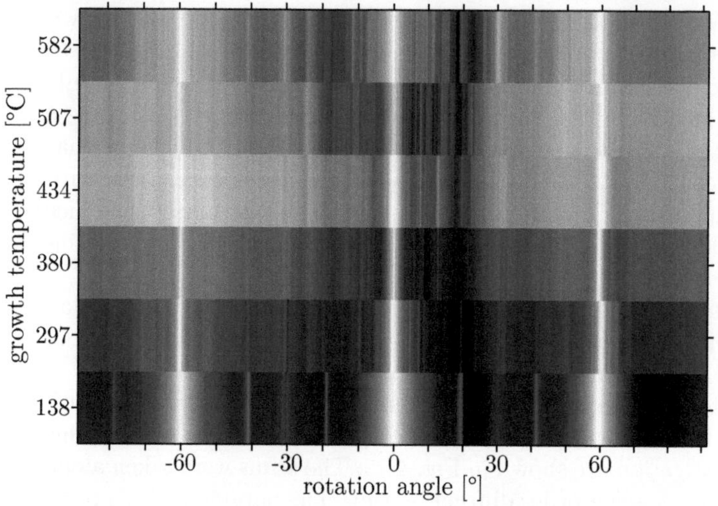

Figure 4.19: Rotational SPA-LEED measurements for various growth temperatures. Bright features correspond to a large coverage on the surface. The small shift in the scans are scanning artifacts since the scan parameters were chosen differently for each scan to optimize the intensity in the majority peaks at $0°$, $\pm 60°$. The deposition rate was adjusted in between the scans for the highest two temperatures to partially compensate for desorption. The scans have been normalized and are background corrected. Afterwards, a Gaussian blurr has been applied to emphasize the main peaks.

to the main peaks at $0°$ and $\pm 60°$. Nevertheless, some rotated peaks can be identified throughout the entire temperature range.

The calculations and SPA-LEED measurements show the same behavior and very similar temperature dependence. Within experimental and simulation limits, both experiment and simulation agree well. Both show similar rotation angles and the observed rotation angles are well described by the calculations. The temperature dependence does not seem to play a large role for both experiment and calculation, while the deposition rate dependence, if not accounted for, will most likely lead to the fact that the intensity in the rotated domains decreases.

Part III

Surface Diffusion and Structure Formation

110

For electromigration experiments, single crystalline islands and islands with large aspect ratios, so called nanowires are used to investigate the effect of electromigration, as also described in Chapter 4. It is known that such islands can be grown in a self organized manner on 4° vicinal Si(0 0 1) surfaces [123]. There, the wires are aligned with the step edges. The formation mechanism of such nanowires is still under debate. Mainly, two mechanisms have been proposed to cause the growth of islands with large aspect ratios. The first mechanism is the strain of the epitaxially grown islands on the substrate which, if the island is large enough, introduces a preferential growth direction, while the island may even shrink in the other direction [124]. The other proposed mechanism is that the islands might grow with such high aspect ratios, due to a very pronounced anisotropy in diffusion [123].

This work presents recent results on the diffusion and diffusion anisotropy of selected metals on Silicon. To estimate the influence of diffusion and diffusion anisotropy and the ES-Barrier, flat and vicinal Si samples have been investigated. Furthermore, the experiments were not only performed for the specific systems and conditions, under which the nanowires grow in a self organized manner, but also for different materials and substrate orientations. In Ch. 5.1, a theory is presented, which was developed on the basis of the experimental observations. Using this theory, it is possible to evaluate diffusion parameters and easily evaluate the diffusion characteristics. On the base of numeric simulations and on the base of facetting and step bunching investigations, a full description of the diffusion anisotropy for vicinal substrates is presented and used to describe the experiments.

The influence of anisotropic diffusion on nanowire formation is investigated in Ch. 8. It can however not exhaustively clear the discussion in the theory where different causes for nanowire growth are discussed [124, 123]. Nevertheless, a new aspect is brought up that might help future understanding of the self organized growth.

Chapter 5

Si(1 1 1)

5.1 Ag/Si(1 1 1)

During deposition of Ag/Si(1 1 1) at temperatures between 500° C
and 690° C it is possible to observe two consecutive contrast
changes in PEEM and LEEM. Figure 5.1 shows an image sequence recorded during the deposition of Ag on Si(1 1 1). The
brightness drops right after deposition is started, beginning at
the step edges (panel (b)). Next, the areas that started to turn
dark now begin to turn bright (panel (d)), until the entire surface is covered by the bright phase. This process shows, how
the different Ag induced reconstructions can be distinguished in
PEEM and are due to the two reconstructions formed at increasing Ag coverages. The darker reconstruction occurs in areas with

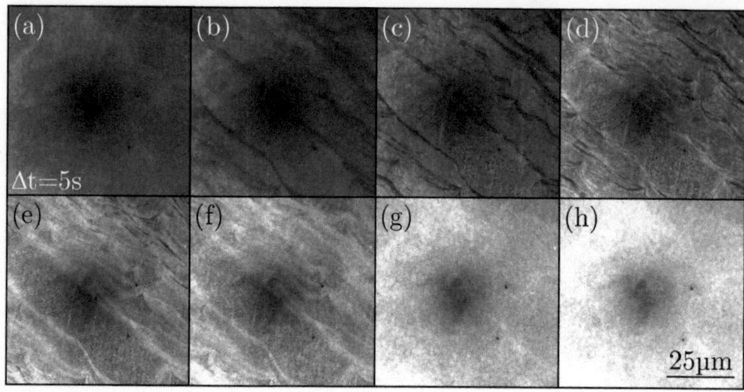

Figure 5.1: Image-sequence of the deposition of Ag on Si(1 1 1). Upon exposure to Ag, step decoration occurs in two steps, starting in panel (b). First, the (3×1) reconstruction (dark) starts to form at the step edges. Later on, starting in panel (d), the $(\sqrt{3} \times \sqrt{3})$ reconstructions starts to form at the step edges (bright), until the entire surface is covered with Ag (panels (g)-(h)).

up to $1/3$ ML coverage [125, 126, 109, 127]. At higher coverages, the $(\sqrt{3} \times \sqrt{3}) - R30°$ reconstruction is formed and fully established at a coverage of 1 ML [110]. The reconstructions start to form at the substrate step edges and as such, the adatoms are preferably deposited at the step edges. This process is referred to as step decoration. Upon further deposition, islands are formed on top of the completely covered surface.

5.1.1   Experimental Observations on Diffusion

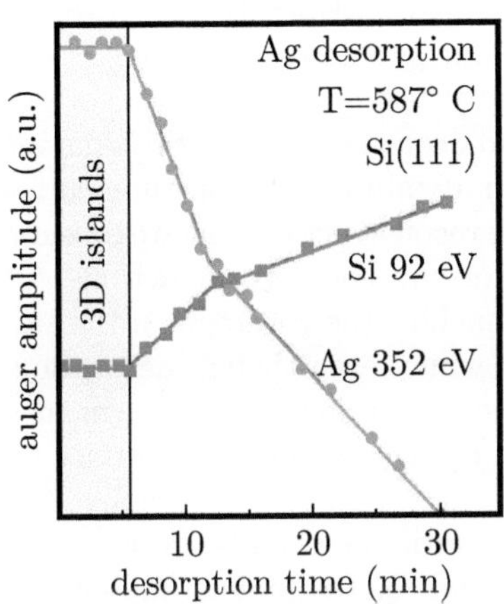

Figure 5.2: Auger Electron Spectroscopy data acquired during desorption of Ag on Si(1 1 1) adopted from [128].

The Ag/Si epitaxial system lends two further advantages for studying surface diffusion: 1) diffusing adatoms of Ag on Si(0 0 1) and Si(1 1 1) constitute an effective 2D system as the 3D epitaxially-grown islands decay in a 2D mode by feeding adatoms onto the surface at the islands' edges; and 2) Ag induces coverage-dependent surface reconstructions at well defined coverages on Si(0 0 1). This was previously described for Si(1 1 1).

When increasing the substrate temperature for desorption experiments, diffusion and desorption occur at increasing rate depending on their activation energy and hopping rate. Figure 5.2 shows an Auger Electron Spectroscopy measurement during the desorption of Ag on Si(1 1 1). Upon the increase in temperature, the deposited silver starts to desorb all over the surface, however the coverage of the surface stays constant for several minutes until the Ag AES signal decreases and the Si signal increases [128]. The kinks at around 12 min desorption time correspond to the change of the surface reconstruction and are linked to the change in desorption rate for the two reconstructions. This measurement shows, that somehow the coverage on the surface remains constant for quite a while, even though desorption takes place. This can be explained by a decay mechanism in which the islands feed adatoms onto the surrounding surface to compensate for the desorbing adatoms and as such keeping the coverage on the surface constant until the islands are used up. This is important to keep in mind for the understanding of the following desorption experiments.

In advance of our desorption experiments, we fabricated homogeneously nucleated islands *in-situ* at elevated temperatures, while observing their growth in real time. At the growth stage where no new nuclei form, we terminated the growth by turning off the deposition flux. We then selected an appropriately isolated island, moved it to the center of the field of view at an appropriate scale for observing its thermal decay; increased the substrate temperature to a value above the growth temperature at which the islands were stable; and digitally recorded the thermal decay of the island. To minimize the time required for the sample temperature to equilibrate, we calibrated the total heating power needed for a particular sample temperature

with an infrared pyrometer before the desorption experiments. Equilibration of the sample temperature to the selected value for observing island decay was observed to take place on a time scale that was negligibly small compared to the islands' decay time.
The desorption experiments in this work all deal with much higher desorption and diffusion rates. Then, the two mechanisms diffusion and desorption lead to a coverage gradient with decreasing coverage from the island perimeter on outward. Once free from the islands the atoms perform random walks. The diffusion length is limited by the average residence time determined by the process of desorption of the atoms into the gas phase. In regions far from islands which do not strongly "feel" the source of adatoms from the island edges, the surface reconstruction begins to break up due to desorption as a result of the increased surface temperature. In these regions the surface reverts back to the substrate reconstruction. Thus, due to the competing processes of diffusion and desorption, as long as the island is able to act as a constant source supplying adatoms to the surface (i.e., before its complete thermal dissociation), the local adsorbate concentration will decrease with distance from a constant value at the island edges. Wherever the Ag concentration falls below the critical value required to form one of the Ag-induced reconstructions (i.e., relatively far from islands) the Si substrate reconstruction will be found, while in the proximity of the islands the Ag-induced reconstructions will be found. It is not difficult, therefore, to conceptualize the break up of the Ag-induced reconstructions far from the islands, and the eventual approach of the boundary, separating the different surface phases, towards

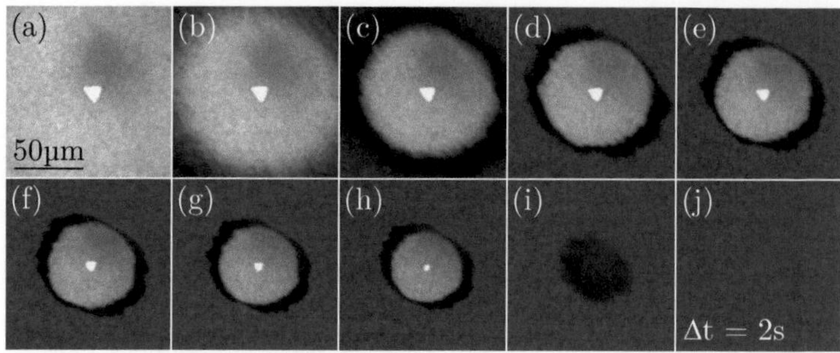

Figure 5.3: PEEM image sequence during desorption at a desorption temperature of $825°C$. The time interval between the images is 2s and the scale is the same in all images. The bright decaying triangle is the single crystalline Ag island, surrounded by a bright $\left(\sqrt{3} \times \sqrt{3}\right) - R30°$ and a dark (3×1) reconstructed area. The outermost area shows the (7×7) reconstruction of the clean Si(111) surface. In images (a)-(c) the iso coverage zones (ICZs) (5.1.2) are forming, while in images (d)-(h) a quasi static equilibrium is reached while the size of the zones only decreases because of the island radius dependence shown in Eq. 5.5. After some time, the island vanishes (i)-(j) and the ICZs can no longer exist since their Ag-atom source, namely the island is depleted. The slight anisotropy in panels (d)-(h) is due to a very small miscut and therefore steps aligned in that direction [129]. The images have been background corrected.

the islands. Interestingly, depending on the particular system, different reconstructed surface phases can produce a different contrast in LEEM/PEEM. Figure 5.3 shows the typical decay of a Ag island during a desorption experiment. After the Ag island reaches the desired size, the Ag flux is turned off and the temperature is rapidly raised to the desired desorption temperature of in this case $825°C$. Panel (a) shows the initial configuration of a bright triangular Ag island surrounded by a $(\sqrt{3} \times \sqrt{3})$ reconstructed background.

During the desorption experiment, in Fig. 5.3, a bright, $(\sqrt{3} \times \sqrt{3})$ reconstructed zone is maintained in the immediate vicinity of the island (panel (b)), but becomes surrounded concentrically by a darker, (3×1) reconstructed region (panel (c)). In panel (c), a second, slightly brighter, (7×7) reconstructed zone, can be observed surrounding the darker (3×1) zone. The specific phases associated with each contrast region were confirmed by microdiffraction (see Fig. 5.4). In Fig. 5.3 (d), two concentric zones of different gray levels, resembling different reconstructions, can be clearly distinguished. During continued desorption, panels (e)-(h), the two so called iso-coverage zones (ICZs) (5.1.2) scale with the island size. Finally, in panel (i), the island's Ag reservoir is depleted and the island disappears, followed by (first) the $(\sqrt{3} \times \sqrt{3})$ zone and (afterwards) the (3×1) zone. In panel (j), all Ag has desorbed, leaving a clean (7×7) reconstructed Si(111) surface behind.

The $(\sqrt{3} \times \sqrt{3})$ and the (3×1) reconstruction were the only Ag-induced reconstructions that we observed during our desorption experiments. The succession of the zones, $(\sqrt{3} \times \sqrt{3})$ in the vicinity of the island, then (3×1), and finally (7×7), shows that the

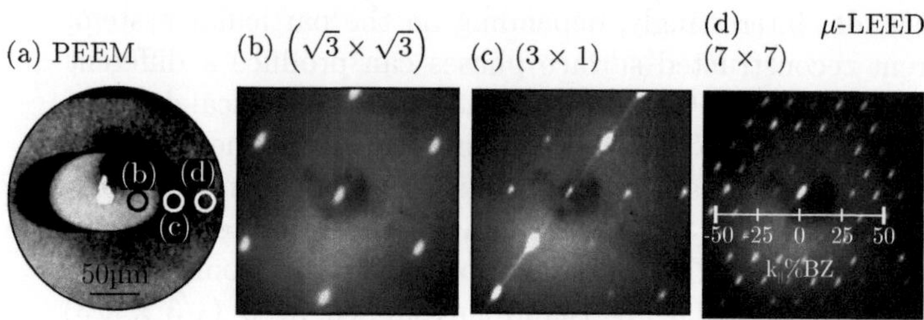

Figure 5.4: Panel (a) shows a PEEM image of an ICZ on a 1° vicinal Si(1 1 1) substrate. The circles indicate the typical areas, where the µ-diffraction patterns (panel (b)-(d)) have been acquired. For experimental reasons, the diffraction patterns were all recorded at the same position, while the different areas passed by during the desorption experiment.

coverage needed for the formation of the (3×1) reconstruction is lower than for the $(\sqrt{3} \times \sqrt{3})$ reconstruction. This observation is consistent with the STM study by Nogami et al. [109], where a (3×1) reconstruction was always reported to be the separating phase between $(\sqrt{3} \times \sqrt{3})$ and (7×7) reconstructed areas. The phase diagrams of Ag on Si(111) as described by Nogami et al. [109] and Le Lay et al. [111] suggest that also a (5×2) [109] reconstruction should be present at some coverage. Neither during growth nor during desorption did we observe this reconstruction in the temperature regime investigated in this study.

Figure 5.4 shows the measured diffraction patterns of the different zones. The anisotropy of the ICZs is discussed in Sec. 7.2.

For the analysis of these zones, a multi-step process is applied to each desorption image-series. As a first step, a particle finding routine with a manual threshold identifies the island as a bright particle and the area of the island is determined. The area is then assumed to be round and a radius of each image of the sequence is thus calculated. As a second and third step, the area of the innermost and second zone is calculated in the same manner as for the island, after which we end up with radii of both zones and the island. All of which are assumed to be round. This method can thus only be applied to desorption experiments on nearly isotropic surfaces (see Ch. 8).

5.1.2   Theoretical description of the ICZs

We now briefly develop a theoretical description of the experimental observations. The theoretical description was developed by I. Lohmar and J. Krug based on our observations. The details of the theoretical model including stability evaluations can be found in [130]. Initially, a single reconstructed zone is formed around an island (see Fig. 5.5 and Fig. 5.3 (b)). The boundary of the zone is given by the location of the critical coverage, necessary to form the inner reconstruction in the coverage gradient. As a first simplification, we assume a rotational symmetry of the problem. Given, that the islands shape does not reflect in the shape of the zones, this assumption seems justified. The speed of the isoline of coverage is given by

$$v_i = \Omega \cdot \left(j_{in} - j_{out} \right), \tag{5.1}$$

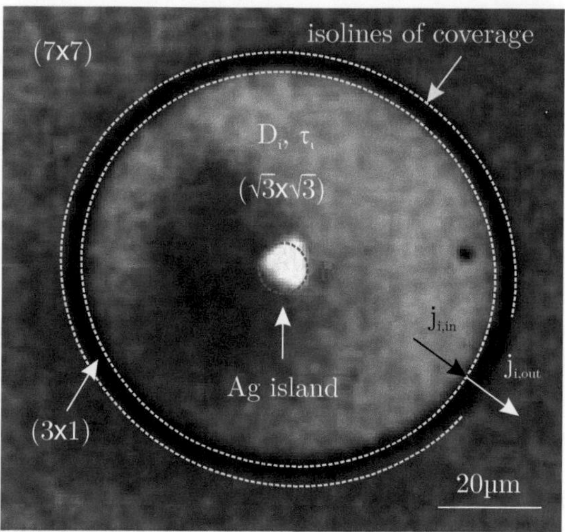

Figure 5.5: Sketch of the relevant parameters and values for an ICZ.

where j_{in} and j_{out} are the flux of adatoms into and out of the boundary respectively and Ω is the space an atom needs on the surface. Therefore, the zones have been named iso-coverage zones (ICZs) [131]. We assume the diffusion to be the limiting factor contrary to the case of Au/Si [132], where the spreading of Gold is limited by the incorporation into the reconstruction. This assumption is however justified by our experimental observations. If the temperature is increased rapidly to very high temperatures, the islands can start moving rapidly similar to water-drops on a hot plate. During this rapid movement, the ICZs follow practically without any lag and the islands are always in the middle of the ICZ. We can therefore assume that the observed situation is well described if we assume any frame of the image series to be a quasi-static picture. We then end up with the steady state condition

$$v_i = 0, \tag{5.2}$$

where both fluxes are equal. This is not limited to a single zone case and we will therefore continue with indexed parameters as defined in Fig. 5.6. If

$$\phi_i = \theta - \theta_i \tag{5.3}$$

is the excess coverage of the $i-1$th reconstruction, the coverage in each reconstructed area can be described by the stationary drift diffusion equation

$$D_i \nabla^2 \phi_i(r) - \frac{\phi_i(r)}{\tau_i} = 0 \tag{5.4}$$

with zone-wise mean desorption time τ_i and diffusion constant D_i (and consequently diffusion length $l_i = \sqrt{D_i \cdot \tau_i}$), derived from Fick's 2nd law [26]. For simplification, we assume, that the desorption time τ_i is the same for all zones. Due to the rotational symmetry, the general solution is given by modified Bessel functions in the appropriate boundary conditions.

In the limit that R_i, $R_{i+1} \ll \ell_i$, and R_{i-1}, $R_i \ll \ell_{i-1}$, in the circular geometry, the steady-state radii are found to obey

$$R_i = R_{i+1} \left(\frac{R_{i-1}}{R_{i+1}} \right)^{\nu_i} \tag{5.5}$$

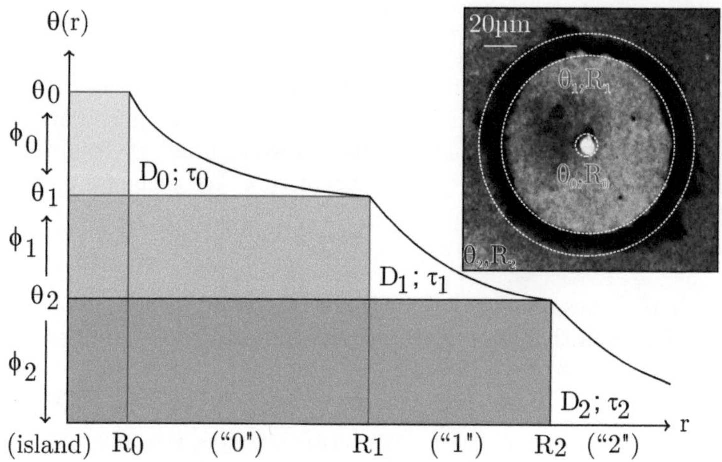

Figure 5.6: Diffusion model for multizone systems with definition of variables and notation.

where $\nu_i = [1 + (D_{i-1}/D_i)(\phi_{i-1}/\phi_i)]^{-1} < 1$ for the i-th ICZ. This reduces the problem of a myriad of diffusing adatoms to a simple power scaling law (Eq. 5.5) that depends only on the diffusion constants and critical coverages of the two adjacent zones. Details of this model can be found in [130]. For the outermost zone, which in our case corresponds to the Si(111)-(7×7) reconstructed area, the outer radius $R_{i+1} \to \infty$, and is replaced by the diffusion length ℓ_i in Eq. 5.5.

The theory presented here suggests that in multi-zone systems, for each ICZ, the adjacent inner zone plays the role of the adatom supply. This result is supported by the experimental results presented in the following section.

5.1.3 Experimental Results for Ag/Si(1 1 1)

In Figure 5.7, the time dependence of the radius of various zones is plotted as a function of time. The innermost zone disappears after the island and the outermost zone disappears after the inner zone. For the Ag/Si(111) system this means that the inner, bright ICZ (supplied by the central island) in turn also feeds the outer, dark ICZ and acts as its adatom supply.

We can now use the general model for multi zone systems to distinguish diffusion parameters on different reconstructions. In Fig. 5.7 (b), the data from Fig. 5.7 (a) is plotted in a double logarithmic scale as shown in the legend. The shown fits represent the power laws according to Eq. 5.5. For the numerical analysis, the area of the triangular island is converted into a circle of the same area as the specific shape of the island does not affect the shape of the zone and the change of the perimeter length is negligible. For the multi-ICZ case presented here, the results varied between 0.04 and 0.11 for ν_1 and between 0.88 and 0.99 for ν_2.

These results on the multi ICZ on Si(1 1 1), with different diffusion constants for different reconstructions, have an interesting implication. First, it is important to consider the critical values for the coverage needed to form the various reconstructions on Si(1 1 1). Although the exact coverages of the involved reconstructions have been under debate for many years, we will assume 2 monolayers to be the critical coverage to form islands, 1 monolayer as the critical coverage for the $\left(\sqrt{3} \times \sqrt{3}\right)$

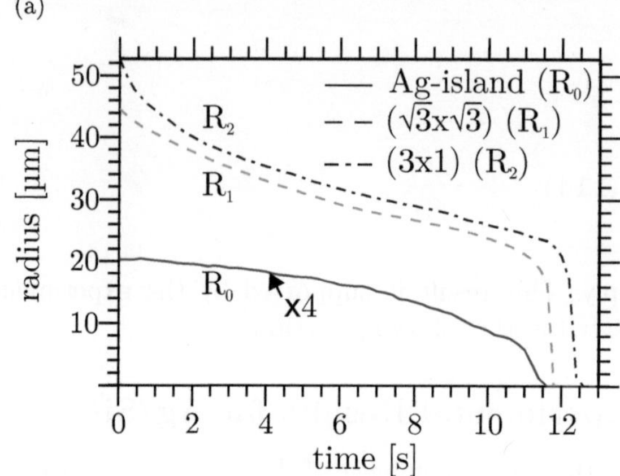

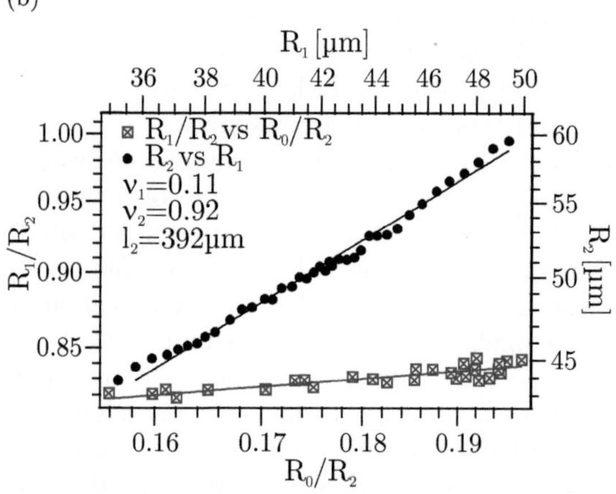

Figure 5.7: Panel (a) shows the typical time dependence during Ag island decay for the island and the two ICZs. After the formation of the ICZ is completed after around 2 seconds, the radius of the inner and outer zone shrinks slowly, similar to the behavior on Si(001) for only one ICZ. Once the island no longer exists, the supply of Ag adatoms disappears and the bright inner ICZ collapses. After the inner ICZ has disappeared, the outer dark zone collapses as well. This time shifted behavior suggests that the inner ICZ is the adatom reservoir for the outer ICZ. The diameter of the island has been enhanced by a factor of 4 for illustration. Panel (b) shows the experimental results of the iso-coverage zone ($R_{1,2}$) and island (R_0) radii for comparison with the theory (see Eq. 5.5) at 870°C. The lines are fitted yielding the displayed values as the only variables. The outermost radius thereby yields the diffusion length ℓ_2 for Ag atoms on the bare Si(111)-(7×7) reconstruction.

reconstruction [110, 133, 109], and 1/3 of a monolayer as the critical coverage of the (3×1) reconstruction [103, 102, 109]. The following argument, however, does not critically depend on the exact values for the coverages since the ν_i's depend linearly on coverage whereas the activation energy enters exponentially. With the typical values of $\nu_1 = 0.08$ and $\nu_2 = 0.95$ and assuming, that the attempt frequency is roughly the same on all reconstructions, this results in the ratios $D_0/D_1 = 17.25$, $D_1/D_2 = 0.026$ and $D_0/D_2 = 0.45$ and illustrates that the diffusion constants are heavily dependent on the underlying reconstruction. We use $E_{D_0} = 0.33\,\mathrm{eV}$ from the literature [32] for the innermost $(\sqrt{3} \times \sqrt{3})$ reconstructed zone. The work by Hanbücken et al. [134] has been performed in a temperature and coverage regime that is comparable to our study. Again, assuming that only the diffusion energies are different for the involved reconstructions, we derive the following activation energies for Ag adatom diffusion: $E_{D_1} \approx 0.61\,\mathrm{eV}$ for the (3×1) reconstruction and $E_{D_2} \approx 0.25\,\mathrm{eV}$ for the (7×7) reconstruction. As the measurements have been performed at various temperatures, several sets of ν_i and D_i with different values have been acquired. Each set, taken for itself leads to the same activation energy with a maximum deviation of $0.07\,\mathrm{eV}$.

It is interesting to compare our value for diffusion on the (7×7) surface with known values from the literature. Jeong et al. have calculated a diffusion barrier of $0.27\,\mathrm{eV}$ for diffusion within a (7×7) half unit cell (HUC) and a barrier of $0.88\,\mathrm{eV}$ for diffusion between HUCs [135]. The latter value was also roughly confirmed experimentally for single Ag adatom diffusion on Si(1 1 1)-(7×7) [136, 137]. However, only the barrier for diffusion within a HUC is consistent with the earlier work by Venables et al. [32] and our work. Apparently, the calculated higher barrier for diffusion across HUCs does not play a role in our experiments: Sobotík et al. [137] stated, in a STM study of diffusion of Ag on Si(1 1 1)-(7 × 7) that "any defect or adatom" within a HUC changes the electronic structure significantly. In our experiment, with coverages between $\theta = 2$ monolayers (close to the island), $\theta = 1/3$ monolayer (at the edge of the (7 × 7)), and $\theta = 0$ monolayer at infinite distances, the situation of several adatoms within a single HUC is a likely scenario. As such, we believe that in our case the diffusion barrier within the (7 × 7) HUC is the relevant parameter. With this assumption, the experiments by Venables et al. [32], our work, and the work by Jeong and Jeong [135] and Sobotík et al. [137] seem to agree.

Our results demonstrate that the imaging of ICZs is not only useful for measuring diffusion parameters, as was shown in our earlier work [131], but imaging of multi-ICZ systems also provides access to diffusion parameters on different reconstructions. In fact, Eq. 5.5 implies a simple relation between the relative widths of different zones in a multi-zone system and the ratios of the corresponding diffusion constants. This can be most easily seen by considering two limiting cases. When $D_i \gg D_{i-1}$, such that diffusion is much faster in zone i than in zone $i-1$, we have $\nu_i \approx 1$ and therefore $R_i \approx R_{i-1}$, which implies that zone $i - 1$ (enclosed by boundaries of radius R_{i-1} and R_i) is much narrower than zone i. On the other hand, $D_i \ll D_{i-1}$, $\nu_i \ll 1$ implies that $R_i \approx R_{i+1}$, and zone i is narrow compared to zone $i - 1$. In general, our model predicts faster diffusion on a reconstruction will yield a wider ICZ for this reconstruction, even though similar effects can be achieved, if the critical coverages are less evenly spaced than in this system.

Chapter 6

Si(0 0 1)

The method we are using to study the surface diffusion on Si(0 0 1) is the same as in the previous chapter and the main difference lies in the involved reconstructions. Here, the Si(0 0 1)-(2×1) reconstruction of the clean surface evolves into the well-studied [138, 139, 140] (2×3) reconstruction upon Ag deposition at elevated temperatures with a critical coverage of 2/3 ML. The indexing and naming of the involved parameters is kept unchanged and can thus be found in the previous chapter. The theory for Si(0 0 1) was already established in the beginning of this work. Nevertheless, this theory is now contained within the new model and is presented in context of the new, more general theory. Some of the experiments presented in this chapter have already been part of my Diploma-Thesis [141] and are only repeated where necessary for the understanding of the following new results.

6.1　Well-oriented Si(0 0 1) Surfaces

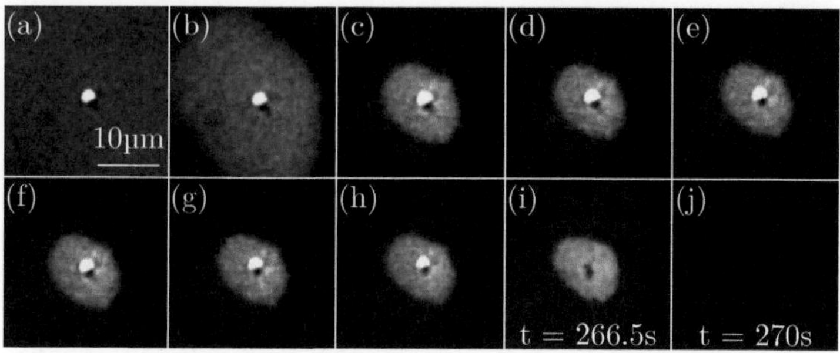

Figure 6.1: Sequence of PEEM images recorded during desorption of a Ag island from Si(0 0 1). The time interval between each adjacent frame from (a) to (h) is $\Delta t = 30$s. The times at which frames (i) and (j) were recorded are indicated in the respective frames. The length scale is the same in all frames.

Figure 6.1 shows a series of images taken during the thermal desorption of Ag from flat Si(0 0 1)-(2 × 1) at 650° C. The brightest feature (i.e., the white dot in the center frames (a)-(h)) is an isolated compact Ag island that was formed during epitaxial growth of Ag on the clean Si(0 0 1)-(2 × 1) substrate at a growth temperature of 600° C. It should be noted that characteristics of the surface morphology remain unchanged at the growth temperature of 600° C after the deposition flux has been switched off. Panel (a) of Fig. 6.1 shows the island immediately after the substrate temperature was increased from the lower growth temperature to 650° C. There is a small rise in the intensity of the surface immediately following the temperature increase, after which the surface begins to break up into dark regions and bright zones that surround each of the Ag islands, as can be seen in panel (b). As time goes on, the bright zone surrounding the island shrinks to a nearly constant size (panels (c)-(h)). In panel (i) the island has disappeared, leaving a smudge that was the surrounding zone, and this smudge itself disappears very shortly after the island has disappeared (panel (j)).

The time dependence of the radii of a typical decaying island and its surrounding ICZ is shown in Fig. 6.2. Three well-defined decay regions are observed for the ICZ: (i) the initial transient decay, (ii) an intermediate steady-state region wherein the zone radius remains essentially constant, and (iii) the final rapid decay region wherein the island and the zone subsequently disappear completely. These regions are not presented for Ag(1 1 1) as they are less pronounced due to the stronger radial dependence of the 2D islands. Consistent with panels (i) and (j) of Fig. 6.1, the bright zone vanishes only after the Ag island at its center has completely disappeared.

To positively identify the nature of the various features in the vicinity of the isolated island during thermal desorption, we exploit the μ-diffraction capabilities of LEEM/PEEM. Figure 6.3 illustrates how microdiffraction is used to determine the structural nature of the bright zones surrounding the decaying Ag islands.

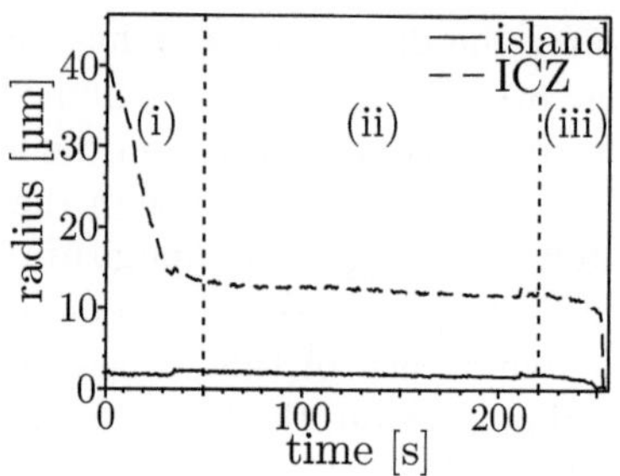

Figure 6.2: The time dependence of the radii of a typical decaying island and its surrounding ICZ, from the beginning of their thermal decay at time $t = 0\,$s to the point where both the island and the ICZ have completely desorbed.

The PEEM image of Fig. 6.3 (a) shows two of the compact Ag islands during thermal desorption at 650° C. To minimize Debye-Waller scattering for the diffraction patterns the sample was rapidly quenched from the desorption temperature of 650° C to a temperature well below the original growth temperature of 600° C. This temperature quenching terminates all surface dynamics. Thus the structural aspects of the surface are "frozen in" nearly instantaneous, producing a snapshot of the surface at various instances during thermal desorption. μ-diffraction patterns of various surface features can then be conveniently produced that represent the exact same structural details as would be found at the higher desorption temperature. While imaging

(a) PEEM (b) (3 × 2) (c) (2 × 1)

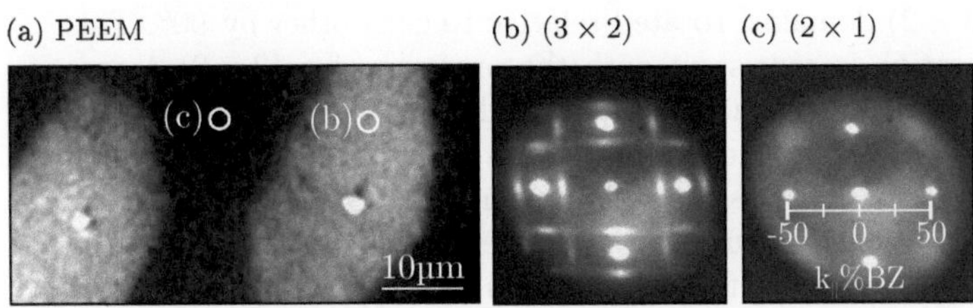

Figure 6.3: (a) A PEEM image of two decaying Ag islands and their surrounding bright zones after the substrate temperature has been increased to 650° C. The two circles labeled (b) and (c) indicate the respective positions on the surface from which the μ-diffraction patterns of panels (b) and (c) were produced. Panel (b) displays a clean Si(0 0 1)-(2 × 1) diffraction pattern, and panel (c) displays the pattern for the (3 × 2)-Ag reconstruction. The reciprocal space scale is the same for panels (b) and (c).

in PEEM, the electron beam can be simultaneously switched on, focused, and deflected onto any feature in the PEEM field of view. The actual electron beam (the image of which is not directly recorded) used to produce the μ-diffraction patterns of Fig. 6.3 was roughly the size of the white circles superimposed on the PEEM image of Fig. 6.3 (a), sufficiently small to completely fit into the different regions of contrast on the surface. With the electron beam positioned at (b) (within the bright zone surrounding the island) the μ-diffraction pattern of Fig. 6.3 (b) was acquired. Figure 6.3 (b) reveals the presence of two identical (3×2) domains, rotated relative to each other by $90°$. This observation is consistent with the presence of a (3×2)-Ag reconstruction on each type of originally Si$(0\,0\,1)$-(2×1) terraces. We thus identify the bright zones surrounding the Ag islands during desorption as the well-known [140, 139, 138] (3×2)-Ag surface reconstruction. With the electron beam located at the position labeled (c) the μ-diffraction pattern of Fig. 6.3 (c) was produced. This pattern displays a reciprocal space pattern of the two identical (2×1) domains of the clean Si$(0\,0\,1)$, representing the dimer rows, characteristic of this surface, that are rotated relative to each other by $90°$ from terrace to adjacent terrace. Thus, the dark regions on the surface in PEEM are bare Si$(0\,0\,1)$-(2×1).

6.1.1   Theoretical Description in the Single-ICZ Case and Results

In the same limits as before (see Sec. 5.1.2), i.e. R_i, $R_{i+1} \ll \ell_i$, and R_{i-1}, $R_i \ll \ell_{i-1}$, in the circular geometry, the steady-state radii are still found to obey

$$R_i = R_{i+1} \left(\frac{R_{i-1}}{R_{i+1}} \right)^{\nu_i} \tag{6.1}$$

where $\nu_i = [1 + (D_{i-1}/D_i)\,(\phi_{i-1}/\phi_i)]^{-1} < 1$ for the i-th ICZ.

For the special case of a single ICZ, the only zone is the outermost zone and the outer radius $R_2 \to \infty$, and is thus replaced by the diffusion length ℓ_2 in Eq. 6.1. We thus find

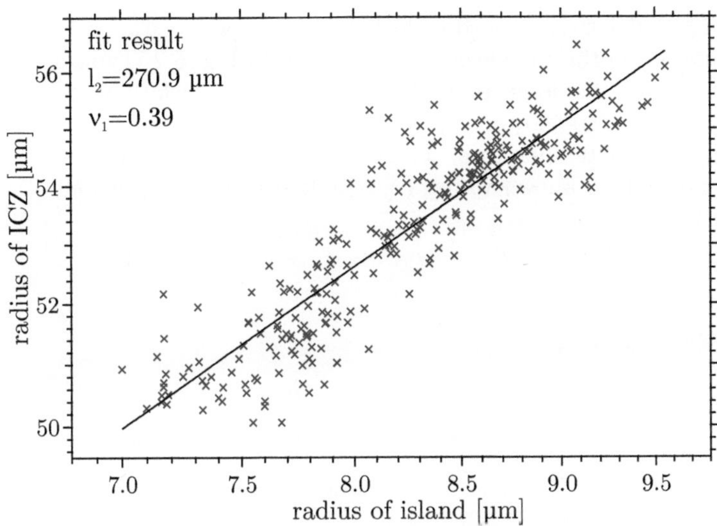

Figure 6.4: Double logarithmic plot showing the ICZ data for Si(0 0 1). The results from the fit using Eq. 6.2 are indicated in the plot.

$$R_1 = \ell_2 \left(\frac{R_0}{\ell_2} \right)^{\nu_1} \tag{6.2}$$

with $\nu_1 = [1 + (D_0/D_1)(\phi_0/\phi_1)]^{-1}$ which is consistent with our previous findings [131], but still relies on the rotational symmetry.

Figure 6.4 shows the double logarithmic plot of the data collected from the time evolution displayed in Fig. 6.2. Typical fit results for the desorption experiments on this surface are in the range of 250-500 µm for ℓ_2 and 0.4-0.7 for ν_1, depending on desorption temperature. These numbers are realistic, and compare well to known values (see Table 6.1 and our previous work).

Physical Quantity	Value	Description	Reference
a	7	mean jump length of a single atom	[142]
E_{D_0}	0.5 eV	activation energy of Ag diffusion inside ICZ	[134]
E_{τ_0}	3.1 eV	activation energy of Ag desorption inside ICZ	[134]
E_{D_1}	0.5 eV	activation energy of Ag diffusion outside ICZ	[142]
E_{τ_1}	3.1 eV	activation energy of Ag desorption outside ICZ	[142]

Table 6.1: Values for diffusion and desorption of Ag/Si(0 0 1).

Chapter 7

The Role of Steps

So far, the presented theory makes use of the rotational symmetry in the diffusion system. If we, however, look at the symmetries at the investigated surfaces, we never find exact rotational symmetry. In the presented cases only multiples of $60°$ or $90°$ symmetry are present. In this chapter, the influence of broken symmetry by single steps and by ordered step arrays on the observed ICZs is investigated.

7.1   Si($0\,0\,1$) with Unordered Steps

It is known from the literature [143, 144] that the adsorption of Ag may influence the step structure. For our experiments, we want to investigate how the changing step structure influences the diffusion fields and with it the ICZs.

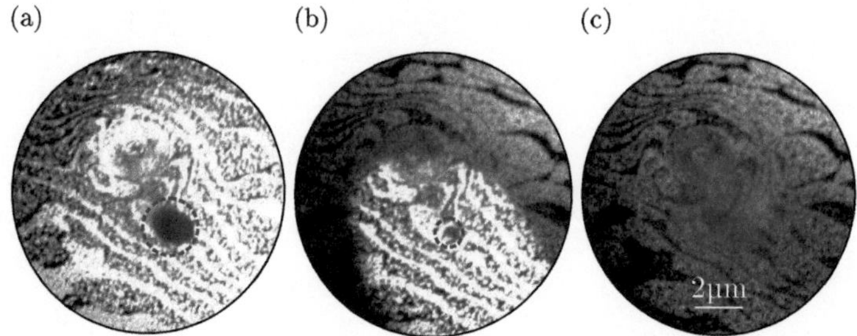

Figure 7.1: LEEM images, in darkfield mode, of a decaying island and its associated ICZ for three successive times. The position of the island in (a) and (b) is indicated by the dashed circle. In (c), both the island and the ICZ have completely desorbed. The substrate step structure is not affected by desorption of the Ag. The length scale is the same in all three images.

We used LEEM in darkfield mode to produce the images of Fig. 7.1. The images were acquired during a typical desorption experiment, this time looking at the sample using the LEEM darkfield mode. In this mode, the involved (3×2) and (2×1) reconstructions are both visible due to the fact that for both reconstructions, two 90° rotated domains are present at the surface with alternating rotation from one terrace to another. For these images, the contrast aperture was placed on one of the 2-fold symmetry diffraction peaks. Therefore, bright terraces are of one rotation and dark terraces are of the 90° rotated dimer orientation. Each boundary of these zones thereby resembles a single height S_A or S_B step. The features associated with the imaging of the (3×2)-Ag reconstruction in dark field are well understood [140], and upon close inspection are reproduced in our images. In each image of Fig. 7.1 the decaying Ag island appears as a dark spot near the center of the field of view, and is highlighted by a dashed circle. In Fig. 7.1 (a) the iso-coverage line is seen to be just entering the field of view in the upper left. The ICZ, consisting of the (3×2)-Ag reconstruction appears bright. In Fig. 7.1 (b) the size of the ICZ has been further reduced. Interestingly, the island also shrinks and appears to scale with the size of the ICZ, but no influence of any single step is visible. That the substrate step structure remains intact has important ramifications for the diffusion model. As such, on macroscopically flat surfaces with low step densities, steps can be neglected in the diffusion model and step associated effects such as the ES-Barrier discussed in Ch. 1.2 do not play a role when it comes to unordered steps. In Fig. 7.1 (c) both the island and ICZ have completely desorbed and the remaining surface reveals the characteristic dark field image of the clean $Si(0 0 1)$-(2×1). A meticulous comparison of the a various regions of the surface before and after the disappearance of the reconstruction reveals an undisturbed step structure as a result of the desorption of the (3×2)-Ag. The only change in the surface morphology is the reversion from the (3×2)-Ag back to the (2×1) reconstruction of the clean substrate. As the step structure does not change, we can neglect the influence of step debunching or similar effects during desorption on low vicinality substrates.

7.2 Si(0 0 1) with Ordered Steps and Step Arrays

The results for surfaces with well ordered steps and step arrays may well be different from the results for unordered steps as the symmetry of the substrate is broken on a macroscopic scale. As described previously, ordered steps can be achieved when 'miscutting' a crystal a few degrees from low index crystal lattice planes. When doing this, however, for Si(0 0 1) surfaces in the [1 1 0] direction, a step ordering and a step height transition have been observed for clean surfaces [145, 93, 92]. When adsorbing metals on these surfaces, the surfaces can undergo massive amounts of reordering and form multiple atomic steps to macroscopic facets [146, 144, 143]. These step bunching characteristics are expected to happen on our surfaces as well and SPA-LEED investigations confirm this fact. However, little is known about the changes in step structure when desorbing the adsorbates especially during our desorption experiments. Before wetreat the associated SPA-LEED investigations, we focus on the associated ICZs and explanations.

For vicinal substrates, the rotational symmetry is no longer present and the presented theoretical results can not be applied as easily. However, results from Ch. 5.1.3 show that narrow zones correspond to slower effective diffusion or at least a higher diffusion barrier. This also is the case for a similar, but fully one-dimensional model [130]. We use this result to understand the results for the anisotropic vicinal surfaces.

(a) Si(0 0 1)-0°   (b) Si(0 0 1)-0.8°   (c) Si(0 0 1)-4°

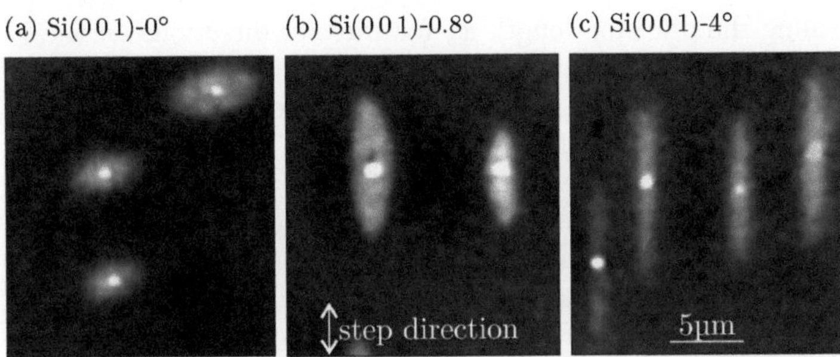

Figure 7.2: PEEM images of desorbing silver islands surrounded by iso-coverage zones (ICZs) [131]. The direction of the step edges is indicated in the centered image. The elongation of the ICZs is always parallel to the step edges, while on well oriented substrates the elongation is due to local variations in step densities. The anisotropy of the ICZs increases with increasing step density (see text). The images were acquired at roughly around 650° C.

Figure 7.2 shows three PEEM images of Ag islands with their respective ICZs during the desorption of Ag from Si surfaces of different vicinality. Panel (a) of Fig. 7.2 shows an almost circular ICZ on a flat Si(0 0 1) surface, while the ICZs of the vicinal surfaces (panels (b) and (c)) exhibit ICZs that are elongated along the step edges of each substrate. As already mentioned, the model cannot be applied to a vicinal surface, and can not easily be modified since the rotational symmetry is no longer valid for anisotropic systems. Instead, while we are unable to determine the diffusion length x_s for the two directions on the surface along the steps and perpendicular to the steps directly, we use the aspect ratio (long axis/short axis of the ICZ) of the ICZs to describe the diffusion anisotropy of the system as a function of the substrates vicinality.

Figure 7.3 shows the ICZ aspect ratios of various vicinal Si(0 0 1) samples with miscuts between $0°$ and $4°$. Each data point represents a single ICZ recorded during desorption at roughly around $680°$ C. For an ideal curve, the temperature has to be exactly the same for all data points as the temperature has an influence on the ICZ shape as will be shown later. For the well oriented surface, the aspect ratio is ≥ 1 by definition and increases gradually with increasing vicinality. It seems to reach its limit at around $4°$ vicinality with an aspect ratio of around 6. The vicinality dependence can also be interpreted as a step-density dependence which has been studied by Pehlke and Tersoff and many others [93, 147, 148, 149, 145] for the vicinal Si(0 0 1) systems. The vicinality dependence of the anisotropy which is presented in Fig. 7.3 can be compared to the plot showing the amount of double steps on a vicinal surface presented by Pehlke and Tersoff in [93]. From a comparison of the vicinalities to the calculated step structure on the surface it can be concluded, that the anisotropy in the system is linked to the amount of double steps on the surface. Even though, the deposition of Ag on a Si(0 0 1) surface can change the step structure on the surface and the occurrence of and the amount of double steps is also influenced by this deposition [149]. It is clear that the step structure should change for example on the $4°$ vicinal surface from step bunches and facets at higher Ag coverages to

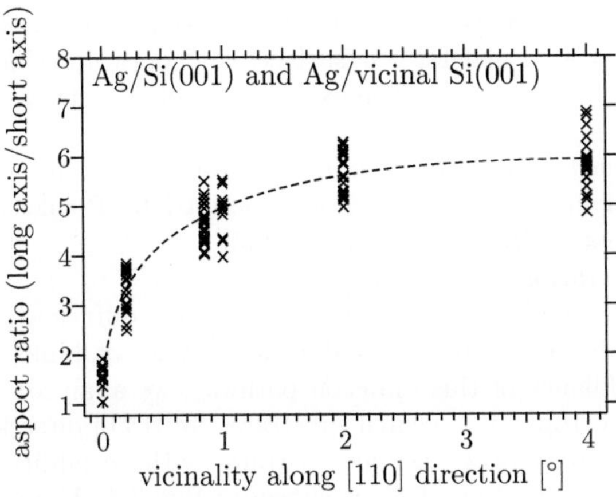

Figure 7.3: The plot shows the aspect ratio of the ICZs for various vicinalities between Si(0 0 1)-0° and 4° misoriented in the [1 1 0] direction. The only reason that the flat 0° samples show aspect ratios slightly higher than 1 is due to local step densities and the fact that the aspect ratio can by definition never be below 1. Starting at the 0.2° misoriented sample the ordered steps on the surfaces lead to strong anisotropies for surface diffusion, since the ICZs only image a footprint for the Ag-adatom diffusion fields. The spread of the data points is due to the fact that these measurements were performed at various temperatures around 680° C to get a significant vicinality dependence even though the temperature plays a role in diffusion anisotropy as is shown in Fig. 7.5. Each data point in the plot was extracted from an individual ICZ.

all double steps on the clean surface. The step structure has, however, been analyzed for flat surfaces and does not change significantly (see. Fig. 7.1). The exact shape of the increase of anisotropy can not solely be explained by this step height transformation. Such a rapid increase in anisotropy does neither reflect in the increase in step density described by Pehlke and Tersoff [93] nor by Tong and Bennett [145].

The observation that the anisotropy is linked to the step density suggests, that the step-edge barrier for diffusion may be bypassed by a possibly lower diffusion barrier at kinks. To check the influence of this diffusion pathway, we analyzed the diffusion anisotropy of 4° vicinal substrates (in [1 1 0] direction) and the anisotropy of 4° vicinal substrates with an additional azimuthal misorientation of 4° as shown in Fig. 7.4. The additional 4° azimuthal introduces rather large kink densities to the well ordered double steps of the 4° vicinal surfaces as sketched in the insets of Fig. 7.4. If the kinks play a role as a diffusion pathway, the anisotropy and with it the aspect ratio of the ICZs should exhibit a strong difference for the two surface orientations. Figure 7.4 shows clearly that the influence of the kink diffusion cannot play a role and can as such be neglected.

The diffusion anisotropy as analyzed during these experiments leads to a ratio of the diffusion constants D in the long and short direction and as such to

$$\frac{D_{long}}{D_{short}} \propto \exp\left(\frac{E_{short} - E_{long}}{k_b T}\right) = \exp\left(\frac{\triangle E_{ani}}{k_b T}\right),$$

with k_b Boltzmann constant, T temperature, E activation energy for diffusion in each direction. The characteristic energy

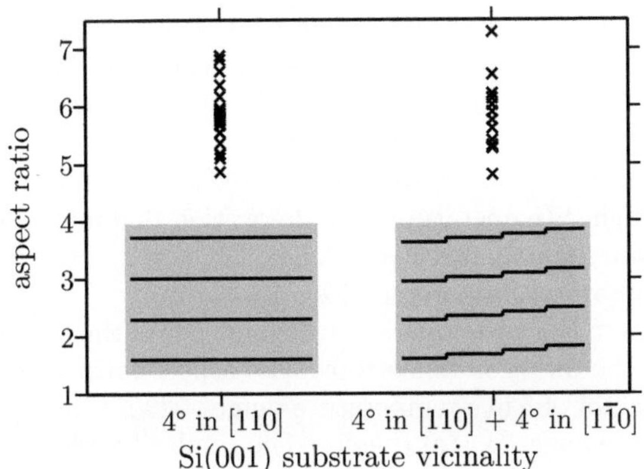

Figure 7.4: The plot shows the aspect ratio of ICZs for two different vicinal Ag/Si(0 0 1) substrates. The 4° misoriented samples show well ordered double steps with very low kink densities on the surface. The 4°+ 4° samples are not only misoriented in the [1 1 0] direction, but also in the azimuthal [1 $\bar{1}$ 0] direction. This results in well ordered double steps, but with high kink densities as sketched in the inset. This should have a strong influence on the aspect ratio, if the introduction of kinks onto a surface leads to new diffusion pathways thus lowering the anisotropy. Thus the plot leads to the conclusion that kink diffusion is not important for the determination of diffusion anisotropy at these typical temperatures of around 680° C.

difference for the anisotropy $\triangle E_{ani}$ should be rather small and may be overcome thermally. Thus we measured the temperature dependence of the anisotropy, the results of which are shown in Fig. 7.5. The temperature range is limited, because at low temperatures, the diffusion dominates and the island density is too high as the diffusion fields overlap and influence one another. This then leads to a (almost) fully Ag covered and reconstructed surface. At higher temperatures the desorption dominates and the Ag reservoir, namely the silver islands, are used up before the ICZs can equilibrate. But nevertheless, the variation of the temperature leads to a clear temperature dependence of the shape of the ICZ, i.e. the diffusion anisotropy. The aspect ratio changes by almost a factor of 5 in a temperature range of $200°$ C as shown in Fig. 7.5 and images (a)-(c) therein. The results are plotted in an Arrhenius type plot to extract the activation energy of the diffusion anisotropy from the temperature dependent measurements. The slope of the linear fit leads to an activation energy of $\sim 0.6\,eV$. This leads to a value for the Ehrlich-Schwoebel barrier [72, 70] for the system $Ag/Si(0\,0\,1)$-$4°$, yet to take this value for granted, one has to be certain, that the two directions decouple and cross talk can be neglected. For large anisotropies this probably is a good assumption, since the anisotropy and the shape of the ICZ does not suggest any cross talk of the long and short direction.

We have presented an easy way to quantify and analyze surfaces for diffusion anisotropy. At least for Ag/Si the anisotropy can be influenced by manipulating local step densities. It seems, that at $2°$ and $4°$ vicinality the aspect ratio of the ICZs is limited by the width of the island. According to this, the anisotropy

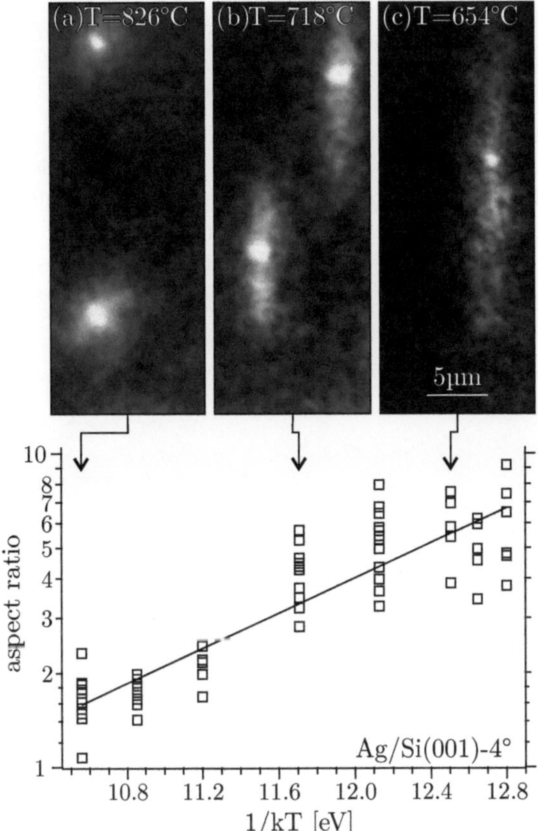

Figure 7.5: The images show typical ICZs during desorption of single crystalline silver islands from a Si(0 0 1)-4° surface, all with the same field of view. The temperature dependence of the aspect ratios of the ICZs is shown in an Arrhenius type plot. The slope m of the linear fit leads to an activation energy in the range of 0.6 eV. This is an effective step edge barrier for the diffusion, maybe even the Ehrlich-Schwoebel barrier [72, 70], even though the influence of changes in step structure may also play a role.

in the system might as well be much larger than it is represented by the aspect ratio. Thus the aspect ratio might have reached its limit at these high anisotropies. We therefore took a numerical approach to justify our assumptions for the anisotropic ICZs and to link the aspect ratio to the diffusion anisotropies.

7.3 Numerical Simulation and Results

The analytical model which uses rotational symmetry cannot be applied to anisotropic islands, we therefore used the stationary drift diffusion equation 5.4 and treated the two perpendicular directions separately. In the simulation, we only treated the simplest case of a single ICZ. The simulation started with a round island with an infinite reservoir in the center of a rectangular lattice and a starting coverage of 1 ML. We simulated for isolated islands as we only analyzed islands which are so far apart from one another, that the influence on the diffusion field was negligible. The influence of this special choice is however negligible as we chose a cell size large enough to yield vanishing cell size dependence. Diffusion from one simulation cell to an adjacent cell was calculated using the anisotropic diffusion constants for the simulation temperature including the different diffusion constants for coverages above and below the critical coverage as well as desorption.

We again started with Fick's 2nd law [26] and derived

$$\frac{\partial \phi_i\left(t, x, y\right)}{\partial t} \;=\; \vec{\nabla}\vec{D}_i\vec{\nabla}\phi_i\left(t, x, y\right) - \frac{\phi\left(t, x, y\right)}{\tau_i} \tag{7.1}$$

$$=\; D_{x,i}\frac{\partial^2 \phi_i\left(t, x, y\right)}{\partial x^2} + D_{y,i}\frac{\partial^2 \phi_i\left(t, x, y\right)}{\partial y^2}$$

$$-\; \frac{\phi_i\left(t, x, y\right)}{\tau_i} \tag{7.2}$$

$$=\; \lim_{\Delta t \to 0} \frac{\phi_i\left(t + \Delta t, x, y\right) - \phi_i\left(t, x, y\right)}{\Delta t} \tag{7.3}$$

with $\vec{D}_i = \begin{pmatrix} D_{x,i} \\ D_{y,i} \end{pmatrix}$. For the numerical simulations [150], using a Taylor series to construct the finite difference approximations to the partial derivatives, we end up with

$$\frac{\phi_i\left(t + \Delta t, x, y\right) - \phi_i\left(t, x, y\right)}{\Delta t}$$

$$\approx \frac{D_{x,i} \cdot \left(\phi_i\left(t, x + \Delta x, y\right) + \phi_i\left(t, x - \Delta x, y\right) - 2 \cdot \phi_i\left(t, x, y\right)\right)}{\Delta x^2}$$

$$+ \frac{D_{y,i} \cdot \left(\phi_i\left(t, x, y + \Delta y\right) + \phi_i\left(t, x, y - \Delta y\right) - 2 \cdot \phi_i\left(t, x, y\right)\right)}{\Delta y^2}$$

$$- \frac{\phi_i\left(t, x, y\right)}{\tau_i}. \tag{7.4}$$

After a few iterations, an equilibrium was formed and the results were quite similar to those of ICZ-PEEM measurements. The iso-coverage line of the critical coverage was identified in the simulation results and its anisotropy was plotted as a function of the ratio of the diffusion constants D_x and D_y, which we put into the simulation. We assumed the same anisotropy (D_x/D_y) for each zone. This is shown in Fig. 7.6. The solid line is a power fit with a resulting power of 0.5. So far it was unclear, if the correct parameter for the anisotropy would be the length of the ICZ − length of the island for each direction or if the diameter of the island would have to be included to calculate the aspect ratio. This simulation, however, yields the same dependence for both cases. As it is easier to simply calculate the aspect ratio without the subtraction of the island, only such experimental data are shown. These simulations could have been carried out for various

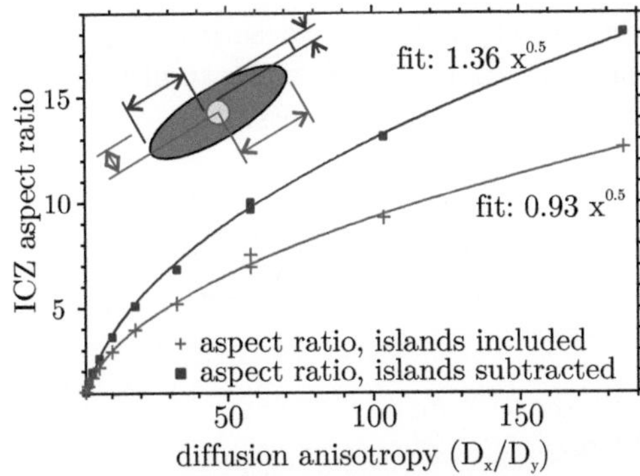

Figure 7.6: Simulation results of the numerical simulation for various ratios of the diffusion constants D_x and D_y. The solid line is the result of a power-fit.

temperatures, the temperature, however, only enters into the calculation of the diffusion constant from an activation energy. Such a temperature dependent calculation is thus obsolete. For different temperatures, the diffusion constants would have to be recalculated and the corresponding aspect ratio can then be found in the plot. This also confirms the results from the temperature dependence of the anisotropy (Fig. 7.5) where the anisotropy decreases for increasing temperatures.

We can now finally compare all aspect ratios to any theoretical prediction using diffusion constants or activation energies as we can convert any aspect ratio to a difference in activation energy for diffusion or vice versa for any given temperature. A dependency of the aspect ratio on the desorption energy has not been found in the simulations and as such can also be neglected for the interpretation of the experiments.

7.4 Facetting and Multi-Step Formation

To investigate the multi-step formation at the specific growth conditions used for the deposition in advance of the desorption experiments, a SPA-LEED UHV-Chamber has been constructed[1], in which the following measurements have been performed. For these measurements, the exact same wafers and therefore vicinalities have been used, that have already been used for the desorption measurements, to yield comparable results. For better comparison, the numbering of the panels is the same for any vicinality in the following three figures. The measurements have been carried out to clear the multi-step formation and facetting issue, that arose in the investigation of the anisotropic diffusion for vicinal $Si(0\,0\,1)$ samples (see Sec. 7.2). For the measurements, each sample was flash-annealed as described previously, then a reciprocal space map (RSM) has been recorded to confirm the miscut and initial step configuration. These scans are shown in Fig. 7.7. All off these RSMs are scaled correctly so that any angle can be analyzed directly from the RSM as the angle in realspace and reciprocal space are equal. The RSMs have also been straightened regarding the $k_\perp$ values, but a straightening with respect to the $k_\parallel$ direction is only possible in 2D scans [151] and could thus not be done for RSMs. The 2-fold symmetry of the (2×1) reconstruction is clearly

---

[1]The SPA-LEED Chamber has been constructed specifically to investigate the facetting and multi-step formation in this chapter, has however also been used for the growth experiments in Chapter 4.3.

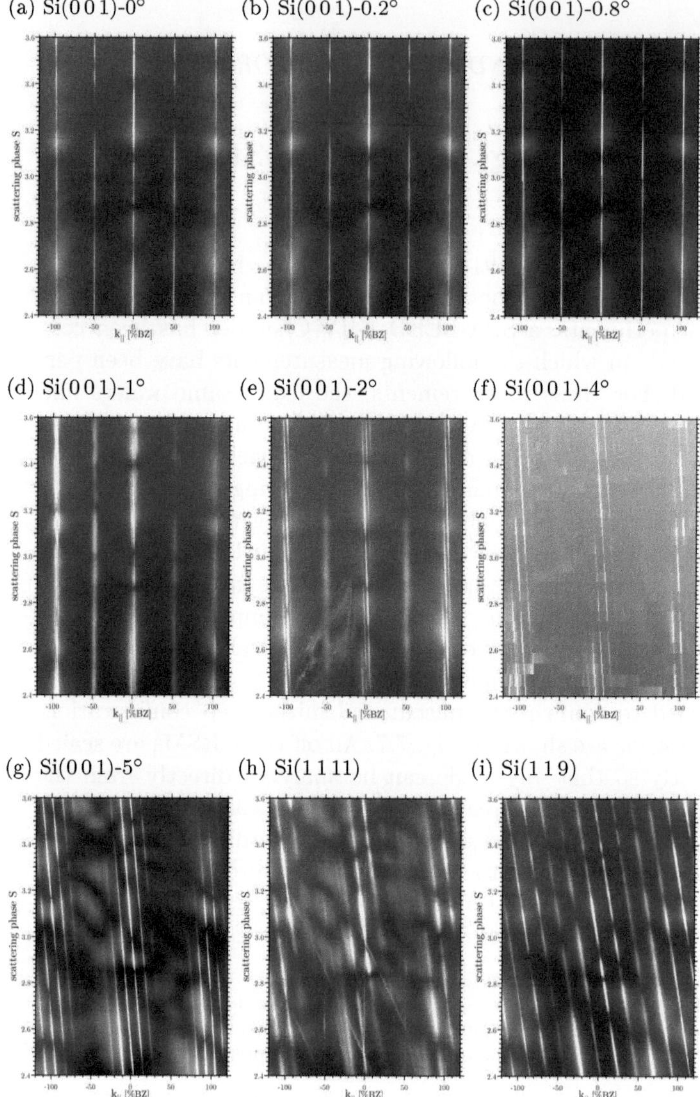

Figure 7.7: Reciprocal space maps (RSM) for various clean vicinal Si surfaces starting with the (0 0 1) orientation as indicated.

visible in the RSMs up to 2° substrate vicinality. In the scans up to 0.8°, no spot splitting is visible, while starting at 1°, the spot splitting due to the regularly stepped surface is clearly visible and increases with increasing vicinality. From about 1° to the Si(1 1 9) surface, the surfaces are double stepped. This could also be confirmed when specifically looking at a single line scan of the RSM for these vicinalities. There, the spot splitting resulting from the regular arrangement of steps can be evaluated and with a known step height can the vicinality be extracted.

After these RSMs have been recorded, the sample was flash annealed several times again before the Ag was deposited under conditions, as close as possible to the preparation conditions in the LEEM instrument. The following RSMs in Fig. 7.8 show the situation after the Ag was deposited and as such at the beginning of the desorption experiments. The low vicinality RSMs of panels (a)-(c) show intensity corresponding to the 3-fold periodicity of the Ag induced (3×2) reconstruction. This intensity is reduced with increasing vicinality until it is practically gone at a vicinality of 1° (panel (d)). This intensity originates in the alternating (3×2) and (2×3) terraces present at flat surfaces. With increasing vicinality, a single to multi-step formation has been predicted [93] and observed [145] previously. During this transformation, one of the possible terraces gains area as the D_B double steps [152] are favored, separating like terraces only and the alternation is therefore disturbed. Due to this transformation, the 3-fold periodicity in the RSMs reduces to 0 as only steps with at least double height are present at a vicinality of 1°. An analysis of the very small spot splitting yields that along with double steps, also 4-fold steps are present, even at

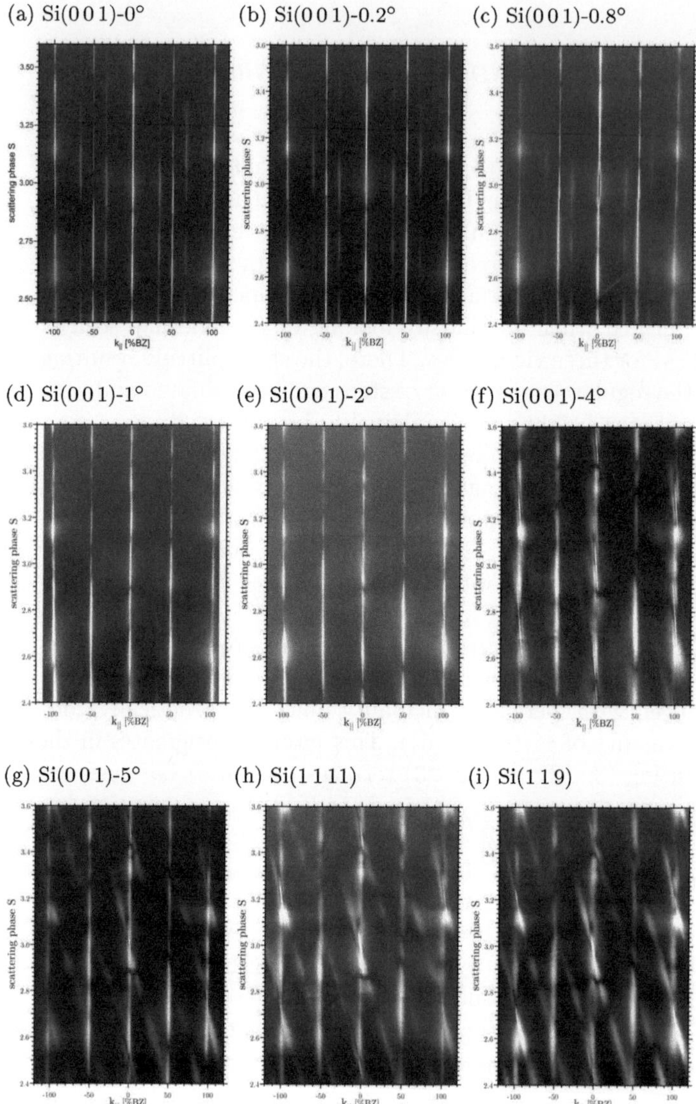

Figure 7.8: Reciprocal space maps (RSM) for various Ag covered vicinal Si(0 0 1) substrates as indicated. Starting with the 5° vicinal surface, facets form upon Ag deposition.

this rather low vicinality. With a further increase in vicinality, a multi-step formation takes place (panels (e)-(f)). For a vicinality of 2° (panel (e)), a mixture of 4- and 6-fold steps are present on the surface, while at the 4° vicinal surface, a mixture of 4-fold, 6-fold and even higher multi-steps are present, which can clearly be seen as the spot splitting is much smaller than for the uncovered surface (compare Fig. 7.7(f) and Fig. 7.8(f)). An analysis of the spot splitting shows the same results. The last three RSMs of the set with vicinalities of 5°, 7.33°, and 8.93° (panels (g)-(i)) all show the same facets 15.8° off the Si(0 0 1) direction, which yields (1 1 5) facets, while the 2-fold periodicity of the (3×2) reconstruction is still present. All of these facets also show a finite size effect, which can be seen when identifying the short sharp rods running through the intersections of the (0 0 1) and facet rods surrounded by missing intensity.

To analyze the change in step structure during the PEEM desorption measurements, the Ag is then desorbed. Again, the conditions have been chosen to be as close as possible to those in the LEEM instrument, but as no chemical contrast or spatial resolution is possible in the SPA-LEED instrument, the desorption conditions have been kept for a much longer time to ensure that all of the Ag has desorbed at the end of the desorption cycle. The last step in this analysis is then the recording of an additional RSM to illustrate the change in step structure, which are shown in Fig. 7.9For a complete analysis, the last three figures have to be compared to one another, which is done in detail in the following paragraphs and summarized in table 7.1.

Panels (a) and (b) show the RSMs of very low miscut surfaces of 0° and 0.2°. Here, the clean surface shows basically the

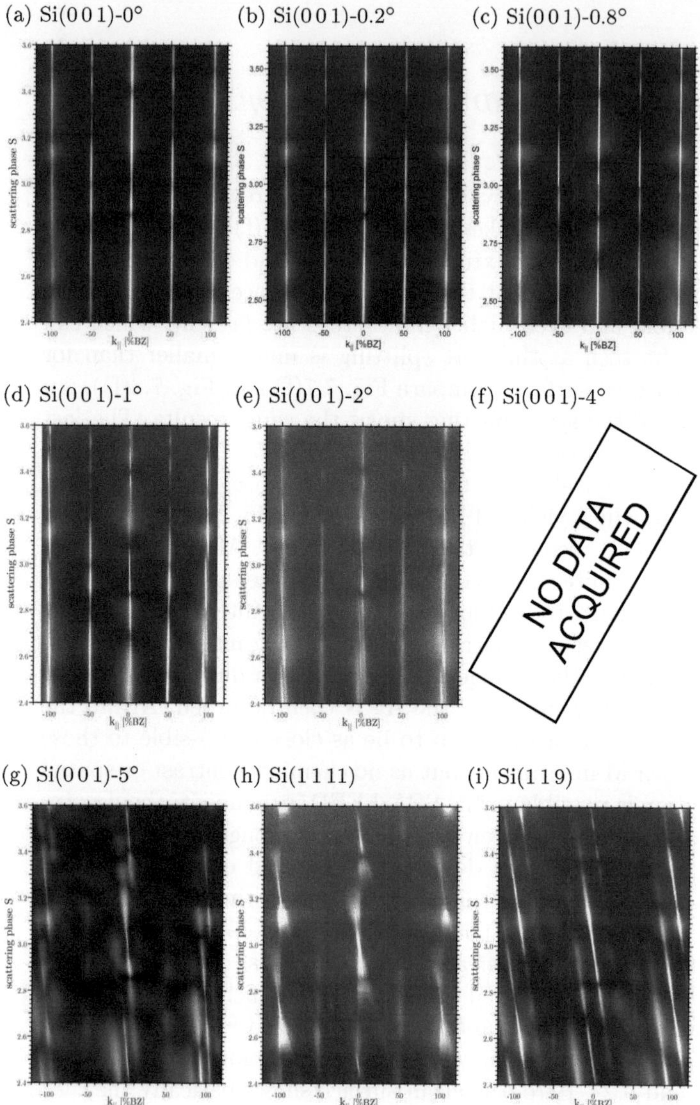

Figure 7.9: Reciprocal space maps (RSM) for various vicinal Si(0 0 1) substrates as indicated. Ag was simply desorbed as it would have happened during a PEEM desorption measurement to illustrate how the step structure evolves during desorption at these specific conditions.

expected pattern for a flat surface without any spot splitting due to regular step trains. Upon Ag deposition, both the 2– and 3–fold periodicity of the two domains of the (3×2) reconstruction can be recognized. These disappear again upon desorption and almost no difference can be seen between the clean surface and the surface where the Ag was only desorbed at the temperatures also used for the PEEM experiments. The measurements on the 0.8° vicinal surfaces are already different. The clean surface does not yet reveal a clearly visible spot splitting due to a regular step train. However, when analyzing single line profiles of these scans, a very small spot splitting can be measured, that fits to the vicinality, if double steps are assumed. Upon deposition, again both 2– and 3–fold periodicity can be recognized showing, that still single steps are present, while the intensity of the 3–fold periodicity is considerably lower than for panels (a) and (b). Upon desorption, the same situation as for the clean surface is reestablished. Beginning with the 1° vicinal surface shown in panel (d), the spot splitting can clearly be recognized in the RSM of the clean surface crossing the $(0\,0)$ rod at each in-phase condition and in the middle in between the in-phase conditions, pinpointing to double steps. Upon Ag deposition, the regular step train disappears again, but the double height steps remain. Here, only the 2–fold periodicity is visible. Therefore, only at least double height steps are present on the surface. Upon desorption, the spot splitting becomes more pronounced again, yet, the ordering seems to be less pronounced. The RSM shown in panel (e) of the 2° vicinal surface shows some charging in the lower left, the structure can, however, still be clearly recognized. The spot splitting is even more pronounced when compared to the 1° surface, still showing 2–fold height steps in a regular step train. The two fold periodicity of the (2×1) reconstruction is less pronounced than for the lower vicinality surfaces, as the dimer covered area gets smaller. Upon Ag deposition, 4–fold height steps form, while the ordering of the regular step train gets lost, most likely due to the size of the Ag induced reconstruction. The 2–fold periodicity, however, becomes more pronounced again. After desorption, the RSM appears similar to that of the clean surface, only that all diffraction peaks are smeared out a lot more, indicating that the ordering is not as good as on the clean surface. The clean 4° surface again shows the double height steps, which turn into at least 4– and 6–fold height steps upon Ag deposition (panel (f)). The ordering of these steps seems almost perfect as even a finite size effect can be recognized for the in-phase conditions. The sample was then taken out to check the island density, thus no desorption experiment was possible. Nevertheless, the island density was very close to that achieved under the similar growth conditions in the PEEM desorption experiments and the results can therefore be compared directly.

The 5° vicinal surface shows the similar regular double step trains as the lower index surfaces in panel (g). However, the ordering of the step train is much better and higher order spots of the regular step train-spot splitting can be recognized. Upon Ag deposition, the surface forms facets with an angle of 15.8° yielding $(1\,1\,5)$ facets. The ordering of these facets is however still very low and the facet rods are therefore smeared out. The 2–fold periodicity of the surface reconstruction is only visible when Ag is present. Upon desorption, the facets disappear again and the regular step trains with double-height form again. The ordering of these step trains is, however, very low compared to the initial surface situation. Most likely, there still are at least some 4–fold height steps left over on the surface as the rods of the spot splitting smear out with increasing distance to the in-phase conditions. This is supported by the fact, that still, the 2–fold periodicity is visible in contrast to the clean surface. The Si(1 1 11) surface again shows double height steps in panel (h), while the ordering is less pronounced than on the 5° vicinal surface. It seems as if many spot splittings are superposed to yield a strange looking high periodicity. After Ag deposition, the surface again reveals the 2–fold periodicity and (1 1 5) facets. This time, the ordering of the facets is much better and again, a finite size effect can be recognized at the intersections of the facet rods and the (0 0) and higher order rods. After desorption, the surface forms several different multi-height steps leading to these strongly dispersing intensity wedges around the in-phase conditions. The RSM in panel (i) of the clean (1 1 9) surface shows the regular structure known for this surface [153]. Upon Ag deposition, the surface again forms (1 1 5) facets and the RSM can hardly be distinguished from that for the Si(1 1 11) surface, also showing the finite size effect. After desorption, the surface looks very different from the initial surface now showing the regular step array of double height steps, this time without any 2–fold periodicity.

Surface	clean	Ag covered	after desorption
0°	single steps	single steps	as on clean surface
0.2°	single steps	single steps	as on clean surface
0.8°	single steps	single & double steps	as on clean surface
1°	double steps, spot splitting	double & four fold steps	less ordered structure than on clean surface
2°	double steps, spot splitting	four & six fold steps	less ordered structure than on clean surface
4°	double steps, spot splitting	four & six fold steps, multi height steps, finite size effect	- no data acquired -
5°	double steps, spot splitting	(115) facets, finite size effect	no facets, multi height steps, less ordered structure than on clean surface
Si(1 1 11)	double steps, spot splitting	(115) facets, finite size effect	multi height steps, strongly dispersing intensity
Si(1 1 9)	double steps, spot splitting	(115) facets, finite size effect	ordered double height steps

Table 7.1: Summary of the results from the reciprocal space maps of vicinal Si(0 0 1) surfaces.

7.4.1 Facetting and Diffusion Anisotropy

Figure 7.10 shows the combined results of all diffusion anisotropy measurements, where the aspect ratio has been converted into a diffusion constant ratio using the numerical simulations. Therefore, each aspect ratio has been converted by the fit results from the numerical simulations: $\frac{D_y}{D_x} = 0.93 \cdot$ aspect ratio$^{0.5}$. The theory by Natori and Godby [69] is then applied, taking the multi-steps into account. Their height has been measured with SPA-LEED and has been extracted from RSMs. The results from all of these techniques and simulation is in combination capable of a proper description of the diffusion anisotropy. None of the presented techniques is taken for itself capable of a correct description of the measured diffusion anisotropy. The combination, however, is capable of a nearly perfect description of the LEEM/PEEM data.

These techniques can also be applied to other surfaces. This will in the following be done for other vicinal substrates with surfaces in between the Si(0 0 1) and Si(1 1 1) orientations.

Figure 7.11 shows the diffusion anisotropy of Ag diffusion for all investigated surfaces. Many of these datapoints except those for Si(1 1 11) and Si(0 0 1)-5° have already been presented in my diploma thesis, but the specifics of the anisotropy and the sudden isotropy for intermediate low index surfaces was not understood at all.

On the basis of this work and the presented combination of techniques, basically any diffusion anisotropy can be understood. From the combination of the presented techniques, we expect an anisotropy as indicated by the dashed lines in Fig. 7.11.

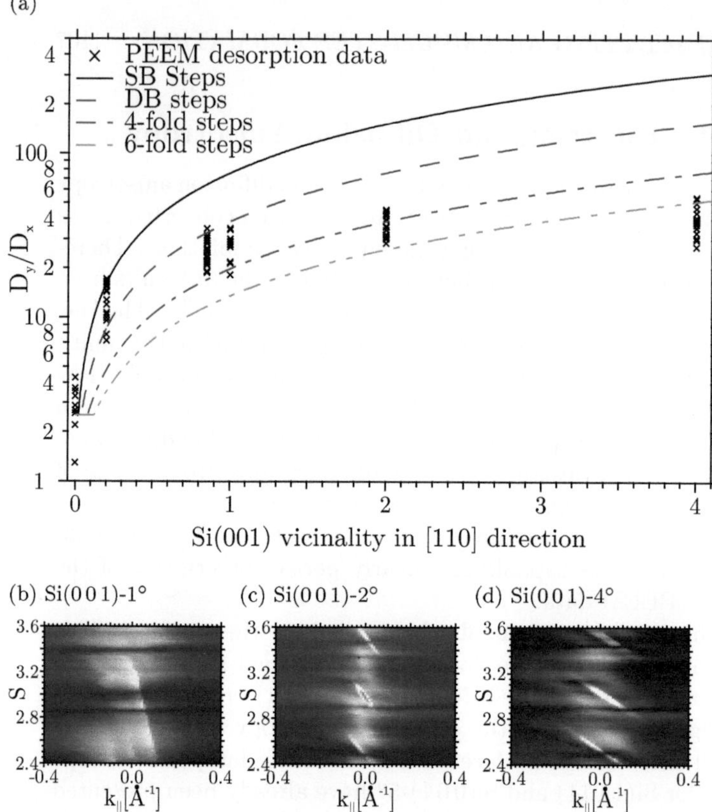

Figure 7.10: The plot (a) shows the diffusion anisotropy measurements in combination with the theory by Natori and Godby [69] and with the step density achieved from the facetting and multi-step SPA-LEED measurements. For better comparison, some of the RSM's are shown in panels (b)-(d) for surfaces with the indicated vicinalities. With the chosen scaling, it is not possible to directly measure angles, but the multi-step character of the surfaces is enhanced. The 1° surface still exhibits double steps, while the 2° vicinal sample already exhibits 4-fold steps. Finally, the 4° surface exhibits 6-fold steps. For comparison, see simulation results for multi-stepped surfaces in Ref. [143].

For the intermediate low-index surfaces Si(1 1 11), Si(1 1 9), Si(1 1 5), and Si(1 1 3), the diffusion is basically isotropic and shows similar values as for flat Si(1 1 1) and Si(0 0 1). This sudden isotropy is caused by a facetting, where the multi height steps come close together and it becomes energetically more favorable to form less steps and therefore to form large facets. To conserve the macroscopic surface orientation, a more or less well ordered array of Si(0 0 1) and Si(1 1 5) facets forms. The size and ordering of this array depends on the macroscopic surface orientation. The transitions from one facet to another will most likely result in a distortion of the periodic surface potential as it is known for single steps and therefore have similar effects as for example the ESB. Transitions are thus likely to result in similar diffusion anisotropy, as single steps only, that the facet-transition-density is much lower than the (multi-) step density. Therefore, the diffusion characteristics of low index surfaces, if faceted are likely to be similar to a flat surface. From the SPA-LEED measurements, it was sadly not possible to extract the size of the facets, even though this is in principle possible.

The SPA-LEED data for Si(0 0 1)-5°, however, show that this surface already shows facets, upon desorption reconverts into a regularly multi-stepped surface which might be the reason for the high anisotropy for a faceted surface. Si(1 1 11) on the other hand redevelops multi-steps, however, not well ordered.

On the right hand side of the graph, the step density increases as the vicinality increases and the diffusion anisotropy accordingly beginning at Si(1 1 1). A multi-step formation as for vicinal Si(0 0 1) was not investigated, but at least a formation of double and even triple height steps is known in the

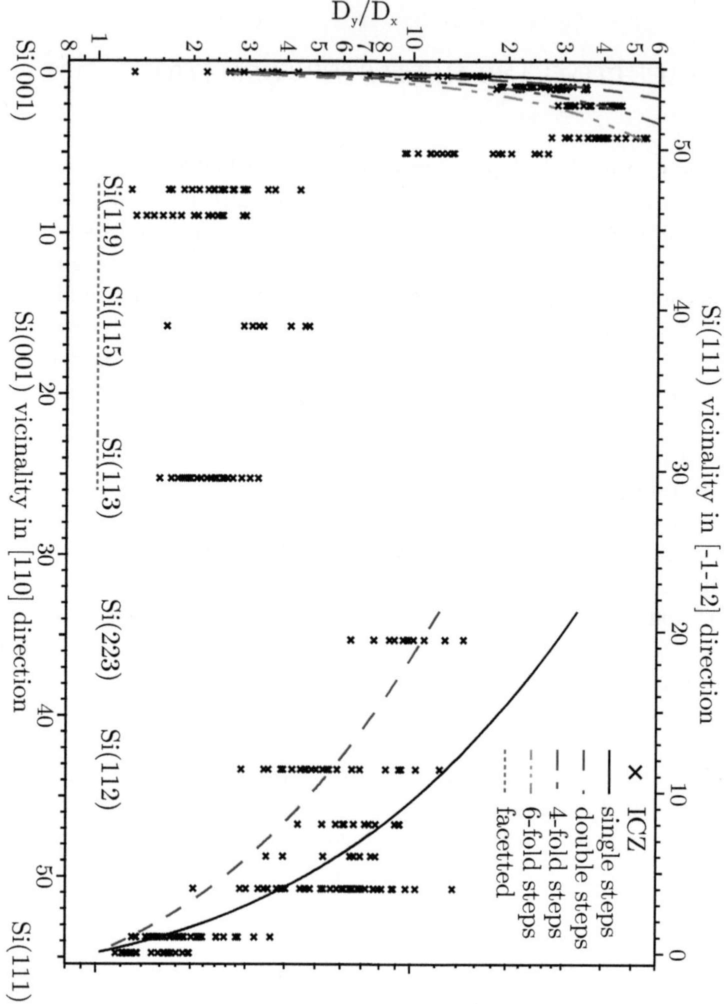

Figure 7.11: Diffusion anisotropy, measured with LEEM/PEEM. The aspect ratio which was shown in previous plots was converted to a diffusion constant ratio using the results of the numerical simulations (Sec. 7.3). The indicated curved lines result from the theory by Natori and Godby [69] and SPA-LEED measurements. Some low-index surfaces have been indicated.

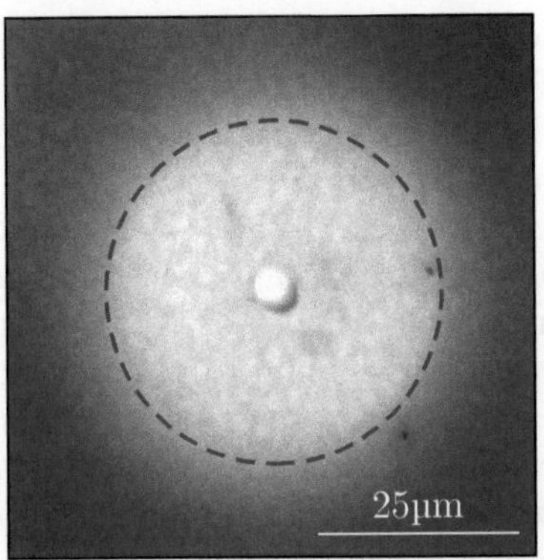

Figure 7.12: ICZ measurement for In/Si(0 0 1)-4° showing an isotropic diffusion zone at a temperature of $T = 730°$ C. The inner ICZ shows the (3×4) reconstruction [154] with a diffuse overlayer. Within this zone, the PEEM contrast is not sufficient to distinguish the diffuse overlayer and the (3×4) reconstruction.

literature [155, 156, 157]. Nevertheless, the data can be described perfectly by the model assuming single and double steps after the aspect ratio has been converted into a diffusion constant ratio. The investigation of the exact step height for vicinal Si(111) surfaces could be part of future investigations. The presented results for the theory by Natori and Godby have been calculated using the presented ES-barrier of $E_S \approx 0.6$ eV and the ledge energy has been assumed to be $E_L = 0$ eV. As no step decoration has been observed during growth pointing to the fact, that the ledge energy is negligible at these temperatures for $Si(0\,0\,1)$. For the theoretical values for vicinal $Si(1\,1\,1)$, the ES-barrier has been measured in a temperature dependent measurement similar to that for $Si(0\,0\,1)$-$4°$ to be in the range of $E_S \approx 0.3\,\text{eV}$ which was then taken for the simulations. Here, step decoration occurs at the growth temperatures and also during desorption, thus a ledge energy of $E_L > 0$ is likely, and $E_L = 0.15\,\text{eV}$ has been (rather randomly) assumed as this fits the data well. The diffusion constants of the flat surface $D_{x,0}$ and $D_{y,0}$ have been assumed to be isotropic and cancel within the diffusion constant ratio. All simulations have been carried out at realistic desorption temperatures for the corresponding PEEM experiment, while the desorption temperatures for vicinal $Si(1\,1\,1)$ up to $Si(2\,2\,3)$ substrates has been higher than those for the experiments on $Si(0\,0\,1)$. Therefore, the results can only be compared in the corresponding regions as the temperature has an effect on the actual diffusion anisotropy. The combination of all the applied techniques is again capable of an almost perfect explanation of the measured data in the entire investigated region. We are hopeful, that this will also be the case for other material systems.

Another attempt has been taken to find another material system which has a similar decay mechanism. $In/Si(0\,0\,1)$ is such a system. For the flat surface, the results are again well explained by our theory. Surprisingly, for the $In/Si(0\,0\,1)$-$4°$ vicinal substrate, the system shows no diffusion anisotropy at all as illustrated in Fig. 7.12.

The involved reconstructions are known from literature and have been confirmed by µ-LEED. Indium is known to induce a

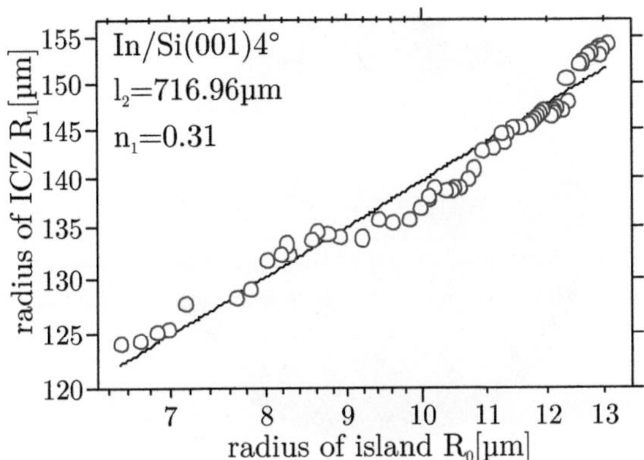

Figure 7.13: Desorption data and corresponding fit using the presented theoretical model shown in a double logarithmic plot. The fit has been carried out using the single zone formula from Eq. 6.2.

(3×4) reconstruction with a diffuse overlayer at our preparation conditions [154]. The PEEM contrast was not sufficient to distinguish the diffuse overlayer and the (3×4) reconstruction during the desorption experiment. The diffuse overlayer is mobile and is similar to any excess coverage in any of the previously presented experiments. We therefore assume that it behaves similar to any excess coverage on the (3×4) reconstruction. We therefore apply the theoretical model for the single zone case to check the applicability. The results are shown in Fig. 7.13. Again, the model describes the data perfectly well and the fit for the single zone case leads to a diffusion length of $\sim 715\,$µm and $\nu \approx 0.3$.

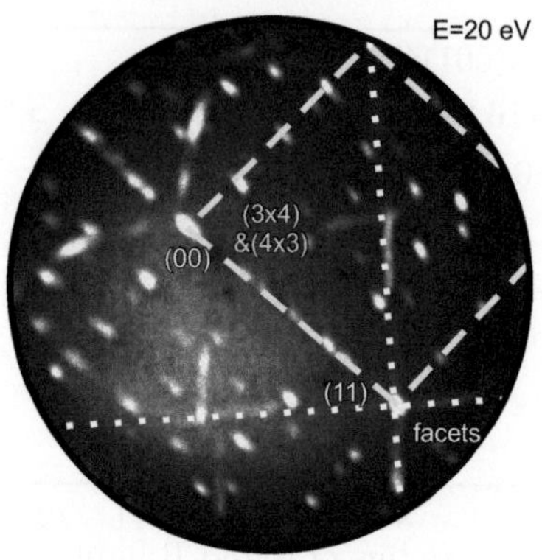

Figure 7.14: LEED image of In/Si(0 0 1)-4° showing the (3×4) reconstruction in 2 domains (yellow, dashed) and the facets [3 1 0] (white, dash-dotted) at an energy of 20 eV.

Similar results have been achieved with less vicinal Si(0 0 1) substrates.

This still leaves us with the question why the steps for this strongly anisotropic substrate does not show any diffusion anisotropy, while it was the most anisotropic substrate for Ag. A close inspection of the LEED image of of Fig. 7.14 shows the 2 domains of the (3×4) reconstruction (indicated by the dashed yellow lines) and the facets (indicated by the dash-dotted white lines). Here, only a quick LEED investigation in the LEEM instrument was performed. Nevertheless, the measurement was sufficient to confirm that these facets run in a direction, close to the $(3\,1\,0)$ direction, which is known to form under our preparation conditions [154]. The results for In diffusion are thus similar to the before presented results for Ag induced facetting and we again would expect isotropic diffusion as it is observed in Fig. 7.12. The closing remark of this chapter can only be that we have come to a full and general understanding of the measurements and the formation of diffusion anisotropy on mesoscopic length scales.

Chapter 8

Diffusion Anisotropy and Wire Formation

Nanowires have attracted significant attention during the recent years. With shrinking electronic devices, the ratio of surface to volume increases, and size dependent material properties dominate the behavior. For example, single crystalline nanowires compared to polycrystalline nanowires, show reversed electromigration behavior [101]. In order to control the growth of self organized single crystalline nanowires, it is necessary to understand at which sites wires nucleate and in which direction the wires grow.

The latter aspect will now be addressed for Ag nanowires on well oriented and vicinal Si(001) substrates. As previously

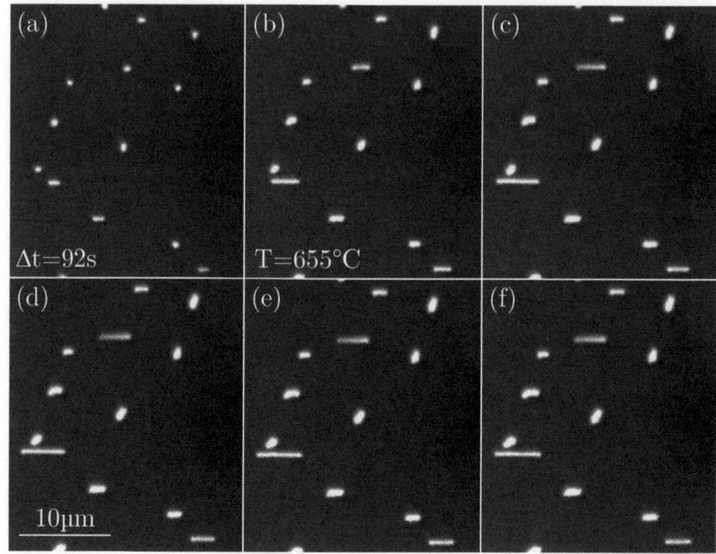

Figure 8.1: Sequence of PEEM images obtained during growth of Ag islands and nanowires on a Si(0 0 1)-4° vicinal surface. (a) nucleation phase, only compact islands are visible. (b) some islands transform into wires. (c)-(f) the wires grow one-dimensionally while the compact islands grow in a 3D fashion. The scalebar is the same for all images.

described, upon deposition of Ag on Si(001), silver islands form on an Ag-induced (3 × 2) [140] reconstruction. These islands are strained due to a lattice mismatch. Simulations [124]

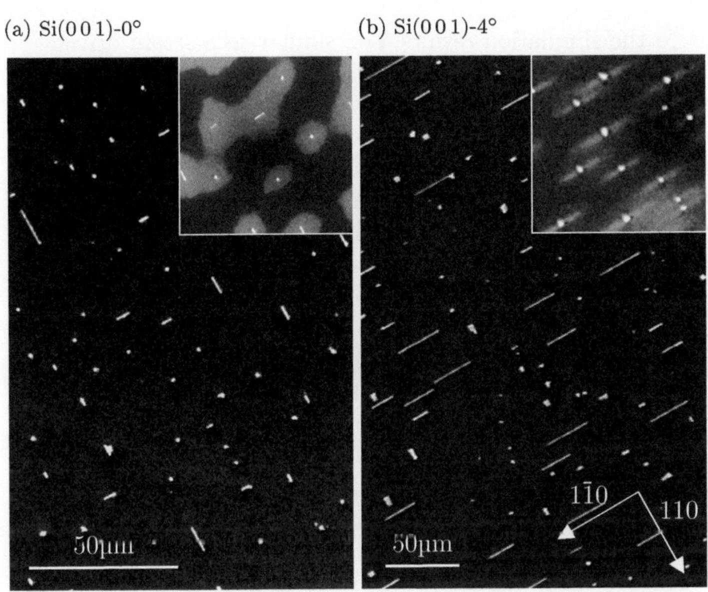

Figure 8.2: SEM images of typical Ag island and Ag nanowire distributions on (a) Si(0 0 1)-0° and (b) Si(0 0 1)-4°. The insets show typical diffusion fields [131] that were acquired with PEEM. The scaling and orientation of the insets and the corresponding SEM images is the same.

predict a transformation of initially symmetrical islands into wires for strained Stranski Krastanov growth. During our growth experiments, we observed [158] shape transitions of Ag islands similar to the simulation results, i.e., similar to a strain driven shape transformation. However, other groups have called the numerical value for the stress of 6% [124] into question, and reported lattice mismatch strains below 0.5% for the same system and similar growth temperatures [159, 160, 161]. Nevertheless, the shape transformation does not necessarily depend on a stress of 6% and should qualitatively still be valid for the observed lower stress. As described in the previous Chapter 7, a considerable amount of diffusion anisotropy is present for Ag on vicinal Si(0 0 1) surfaces that might also favor wire formation [123]. Here, we combine a statistical analysis of the wire formation with the data for the diffusion anisotropy, adding another piece of vital information on the way to a full understanding of the different influences on the wire formation.

The experiments were performed in the LEEM instrument. The same well oriented and vicinal Si(0 0 1) substrates have been used as before. The Ag deposition took place at elevated substrate temperatures, this time in the range of 500° C-750° C, and was monitored *in-situ* with PEEM or LEEM. LEEM darkfield contrast [162] was used to analyze the influence of steps on the nanowire formation. To gain statistically significant data on the orientation of the wires, the samples were also analyzed *ex-situ* with SEM.

As previously described, the Ag adatom concentration gradually increases during deposition, leading to a phase transition from the initial (2 × 1) to the Ag induced (3 × 2) reconstruction [140]. Once this phase completely covers the surface, 3D crystalline islands and single crystalline nanowires form atop this intermediate layer. Figure 8.1 shows an image sequence of a typical growth experiment on a 4° vicinal Si(0 0 1) surface. In Figure 8.1 (a), all Ag islands have a similar shape and it is impossible to tell which island will eventually turn into a wire. When the islands have reached a certain size (Fig. 8.1 (b)), some of the islands become elongated. Afterwards, the wires grow only in the elongated direction while the compact islands continue their normal 3D growth (Fig. 8.1 (c)-(f)). On Si(0 0 1), nanowires form in both $[1\,1\,0]$ and $[1\,\bar{1}\,0]$ directions as can be seen in the SEM overview image shown in Fig. 8.2 (a). In contrast, on the Si(0 0 1)-4° vicinal surface, the nanowires form only in one of the two symmetry directions on the surface (Fig. 8.2 (b)), namely, parallel to the double steps. The insets show PEEM images of the corresponding ICZs and therefore characteristic diffusion footprints. For Ag(0 0 1) islands, which are forming above 600° C [161, 160], the lattice mismatch is identical in the two principal lattice directions, and the wires are evenly distributed along the two possible dimer directions. Although, the lattice mismatch of a Ag(0 0 1) island on a Si(0 0 1) substrate will be the same as the lattice mismatch for a Ag(0 0 1) island on a vicinal substrate, the distribution of the wires changes significantly. On a 4° vicinal surface, the wires are all aligned to the step edges. We believe this change in the wire distribution to be induced by diffusion anisotropy. In our previous work [131] and the preceding Chapter 6, we established an imaging technique to observe diffusion fields in PEEM and a complete description of the observed anisotropic diffusion is presented. As already described, arises a contrast on Si(0 0 1) and different reconstructions can be imaged as bright zones during island decay (see

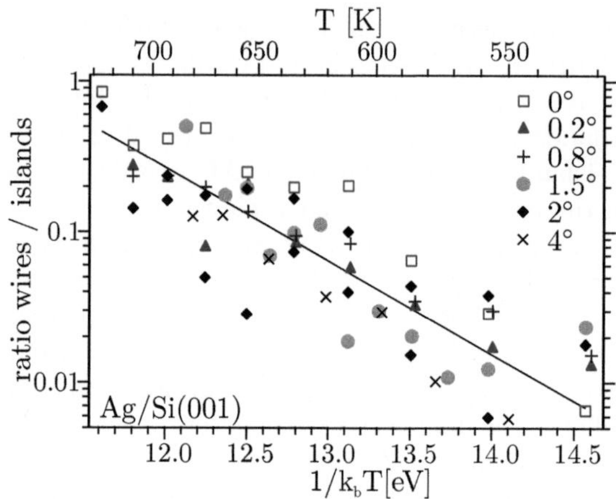

Figure 8.3: Temperature dependence of the ratio of wires to islands for all substrate vicinalities investigated. The plotted ratio of wires/islands is independent of the substrate vicinality. The slope of the line fit in this Arrhenius type plot yields a mean activation energy of 1.4 eV, independent of substrate vicinality.

insets in Fig. 8.2). If anisotropic diffusion is present, as in the case of Ag on vicinal $Si(0\,0\,1)$, these zones are elongated in the direction of lower diffusion barrier, i.e., along the step edges in the direction of the wires in the inset of Fig. 8.2 (b).

For the wire formation, many parameters are relevant. Particularly important is the temperature, as the formation of Ag nanowires on flat and vicinal $Si(0\,0\,1)$ substrates is temperature dependent. When the temperature from one growth experiment to another is increased, the percentage of islands evolving into wires increases as well. Figure 8.3 shows the temperature dependence of the ratio of wires to islands. For this plot, approximately a total of 1000 islands and wires have been investigated to give sufficiently reliable data. The Arrhenius type graph unambiguously shows that the wire formation is thermally activated. The activation energy to form a wire is 1.4 eV higher than the activation energy needed to form a compact Ag island and is independent of the substrate's vicinality. In the investigated temperature range, the nucleation density changes dramatically for a constant flux. To eliminate any possible influences of the deposition rate on the probability of wire formation, we adjusted the deposition rate to yield a constant nucleation density for all experiments in Fig. 8.3. Furthermore, we investigated how the change in deposition rate would reflect on the ratio of wires/islands in Fig. 8.3. Surprisingly, for each temperature and vicinality investigated, the ratio of wires/islands was independent of the deposition rate, nucleation density and vicinality.

Previous work [124] already reported that the wires overgrow the steps of the well oriented Si(0 0 1) surface, and we can confirm that observation. On well oriented Si(0 0 1) samples, wires are oriented randomly to the locally corrugated step edges, while they can overgrow many terraces. As can bee seen in Fig. 8.4. The wires appear dark in all panels. Panels (a)-(b) and (c)-(d) have each been recorded on the same Si(0 0 1)-0° sample. The pairs of images show wires that have overgrown several steps. The pairs of wires shown in (a)-(b) and (c)-(d) have grown in perpendicular direction, yet parallel to one of the dimer rows of the underlying substrate as has been confirmed by µ-LEED.

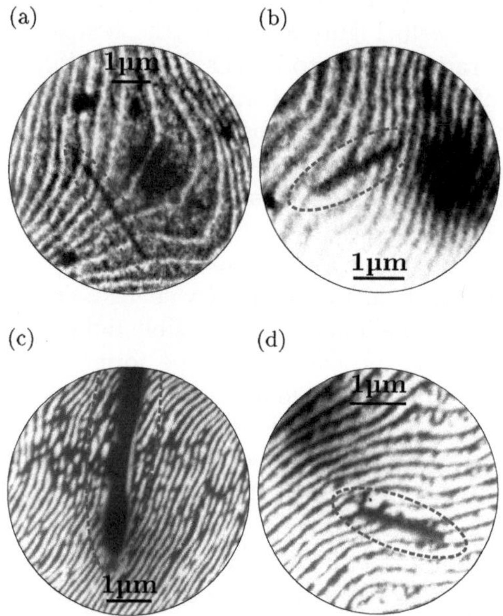

Figure 8.4: Darkfield LEEM images recorded on slightly vicinal Si(0 0 1) substrates after deposition. The images (a) and (b) have been recorded on the same sample showing perpendicular wire directions. The wire in (a) is almost perpendicular to the step edges. The images (c) and (d) have been recorded on a second Si(0 0 1)-0° sample, again showing wires in perpendicular directions, while the wires in all cases overgrow several step edges.

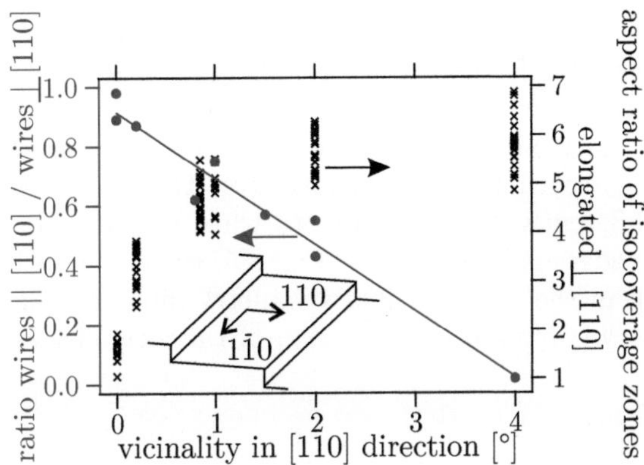

Figure 8.5: Vicinality dependence of the wire direction and the diffusion anisotropy. The solid line (left axis) shows the ratio of wires perpendicular and parallel to the steps. At higher vicinality, the wires align with the step edges. The dashed line (right axis) shows the aspect ratio of the diffusion field. The diffusion anisotropy increases with increased substrate vicinality [129]. For the specifics of the increase of the anisotropy see Fig. 7.10.

It would have been possible that the wires while growing push the steps together to form step bunches around them, this is however not the case for these wires. They do not have much of an influence on the underlying step structure.

In contrast, on the 4° vicinal substrate, practically all wires form along the [1 1̄ 0] direction parallel to the step edges.

The black crosses in Fig. 8.5 show the aspect ratio of the iso-coverage zones [131] for substrates of different vicinality [129]. The specifics of the described increase has been discussed in the previous Sec. 7.4.1. Single crystalline nanowires are formed on well oriented and vicinal Si(0 0 1) substrates. The gradual transition from differently oriented wires to wires that are aligned with the step edges is illustrated in Fig. 8.5, where the ratio of wires growing in the two principal symmetry directions of the Si(0 0 1) surface is plotted as a function of the vicinality (solid line).

Part IV

Closing Remarks

Chapter 9

Conclusions and Outlook

9.1 Island Formation

The self organized growth of Ag islands on Si(1 1 1) is studied as a function of temperature, while the flux remained unchanged throughout these experiments. With increasing temperature, the shape becomes increasingly more regular. The triangular shape is found to be the equilibrium shape of the Ag(1 1 1) islands as this is the shape that remains at high temperatures when most other kinetic limits are overcome thermally.

The crystalline islands show angles of 60°, 90°, and 120° even though the crystallinity of the substrate would only yield

$60°$ and $120°$ angles. Selected area LEED (μ-diffraction) was performed to analyze the islands which even showed a strong contrast in PEEM mode. It was found, that the Ag islands consist of areas showing both $(0\,0\,1)$ and $(1\,1\,1)$ orientation. Despite the lack of atomic resolution, it was possible to confirm that both areas are separated by a grain boundary in the $[1\,\bar{1}\,0]$ direction. The $(0\,0\,1)$ areas can be converted into $(1\,1\,1)$ orientation either by extensive growth or a sudden increase in temperature. The converted areas then show the same orientation as the attaching $(1\,1\,1)$ areas. We can only speculate about the origin of such a transition, it is however likely that the transformation is triggered by strain effects. This is the only origin we find likely. Only one of the most common two contact lines in between the two areas shows a low-index-commensurate structure, any additional length, for instance caused by growth or a temperature difference causes additional strain. Therefore, a transition is likely. Since we never observed freestanding $Ag(0\,0\,1)$ islands, a starting point for the transition is always present and a layer-wise transition will have very low energy cost to it.

In addition to the two different crystallographic orientations, many more rotation angles have been observed by LEED, PEEM and SEM with respect to the substrate orientation. To explain the possible rotation angles, a coincidence site lattice model approach has been taken. While CSL theory is more commonly used in homoepitaxy to explain grain boundaries, only few attempts have been made to explain heteroepitaxial systems. For heteroepitaxy, the occurrence of perfect coincidence is almost impossible and usually coincidence is assumed within a certain distance of the lattice points. This attempt has also been taken and leads to the classical CSL parameter Σ. This parameter does however lead to bar-diagrams where the width of each bar is caused by the assumed coincidence distance. To improve this method for our system, a slightly different approach has been taken in addition. We assumed that the coincidence will also be influenced by this coincidence distance and we therefore calculate a quality factor Q which is given by the binding distances of any lattice site to all lattice sites in the other lattice. This parameter is then used as a measure of good overlap.

The experimental data to the rotation angles of the islands with respect to the substrate can almost perfectly be explained by the quality factor Q. For comparison, we only fitted the experimental data with multi-Lorentzian fits where the only fixed parameter is the position of each fit which was taken from the calculations of Q. The quality factor Q is therefore also a good measure for the description of rotated lattices and due to the better resolution of the data, is an improvement to classical CSL theory.

Upon a rapid increase in temperature after island growth, with flux turned off, it is possible to grow holes in the center of islands. It was shown with μ-diffraction, that the hole showed the Ag induced Si(1 1 1) surface reconstruction and thus reaches right down to the substrate. With turned on flux, a situation can be achieved where the outside of the islands remains almost constant, while the islands' hole grows. Therefore, the ideal shape would be a very thin ring as a remainder after this.

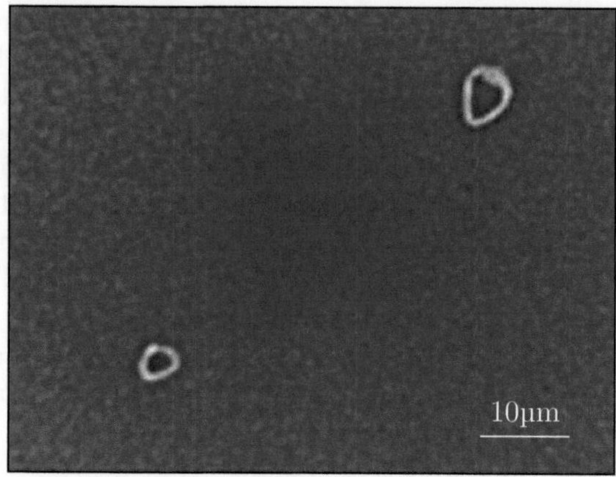

Figure 9.1: PEEM image of a two ring shaped island, where the hole was grown so large that only a thin Ag line remains.

First attempts to grow such rings have been successful and a PEEM image of such a ring is shown in Fig. 9.1. Such rings might be used to study surface plasmon polaritons. Here, the coupling into the island is possible at any incidence direction and the plasmons might be totally reflected on the island to run in circles around the hole. Then, the effects of some particle deliberately placed inside this single crystalline ring would be very interesting to be studied as basically nothing is known about what would happen. Furthermore, if the hole is even grown larger, the ring will eventually rip apart and have thin edges at its opening. At these edges, it might be possible to trigger nonlinear processes when illuminating with a laser due to the tip induced field enhancement at the tips.

9.2 Diffusion and Diffusion Anisotropy

Various techniques have been used to study surface diffusion in the past. All of these techniques have advantages and disadvantages to them. For most techniques, it is hard to tackle diffusion anisotropy and only very few techniques exist to investigate diffusion anisotropy. The most commonly used are adatom tracker or video STM limited to very low temperatures and SEM/Scanning Auger Electron Microscopy investigations on pre-deposited patches, deposited *ex-situ* through masks. Here, we use a simple PEEM or LEEM desorption experiment to investigate surface diffusion [131] *in-situ*. We previously developed a model that explains the findings for Ag/flat Si(0 0 1). We present new results for the system Ag/Si(1 1 1). The known model was not sufficient to explain these results as it was not able to deal with more than one ICZ. Therefore, we present an extended model developed in cooperation by Lohmar [130]. Some of the assumptions that are done in this model are: First, the desorption is assumed to be the same on all reconstructions. Second, we assume, that the image represent a quasi-steady state, which is justified by the experimental observations. Third, only the excess coverage on each zone is assumed mobile. Fourth, radii of the zones are assumed to be much smaller than the corresponding diffusion length. Fifth, the steady state is given by the flux into the zone boundary and out of it in contrary to the situation observed by Ferralis et al. [132] where the results are dominated by the incorporation time into the reconstruction.

The extended model shows that each zone acts as an adatom source for the neighboring outer zone, just like the island is the adatom source for the innermost zone. We have shown that the relative widths of the zones basically reflect the diffusion constants relative to the other zones (see Eq. 5.5), and that the diffusion parameter can vary dramatically between different reconstructions. For the Ag/Si(1 1 1) surface, we find the highest diffusion barrier E_{D_1} of ≈ 0.61 eV in the dark (3×1) reconstructed region, sandwiched between a higher coverage $(\sqrt{3} \times \sqrt{3})$ phase and the (7×7) phase of the bare Si(1 1 1) surface. Considering E_{D_0} of 0.33 eV on the $(\sqrt{3} \times \sqrt{3})$ reconstruction [32] and $E_{D_2} = 0.25$ eV on the (7×7) reconstruction, we conclude that the diffusion of an Ag adatom expelled by the island starts out fast, slows down in the (3×1) region and speeds up again in the (7×7) region. This finding is directly reflected in the widths of the zones relative to each other, keeping in mind, that different coverage ratios have a similar influence. Imaging of ICZs in a multi-zone system is thus not only capable of acquiring diffusion parameters or diffusion anisotropy of the investigated system, but also provides an estimate for the diffusion barriers on different reconstructions in a very quick and simple experiment. Furthermore, with increasing number of zones, this technique can also visualize diffusion gradients on the surface, since the critical coverage for each reconstruction is either known or can be determined in a simple growth experiment.

For ICZs on Si(0 0 1), nothing is changed by the new model as our 'older' model [131] is now contained within the new theory and is the special case with only one ICZ. The increase of anisotropy with increasing steps is due to an Ehrlich-Schwoebel type barrier [72, 70] present in the system. The results from the simple diffusion model presented in Ref. [131] used to explain the diffusion zones can only be used for circular zones. Thus the aspect ratio of the ICZs only presents a number to quantify the diffusion anisotropy of the system. The influence of kink diffusion as a bypass mechanism for diffusion across steps does not influence the anisotropy of the system Ag/Si at these high temperatures (600° C-850° C). Temperature dependent measurements lead to an activation energy in the range of $0.6\,eV$, which is assigned to an effective Ehrlich-Schwoebel barrier of the system Ag/Si(0 0 1)-4°. So far, all of the presented experiments yield data that are consistent with those known in literature, even though those have been gathered with various different techniques.

As we now know the height of the Ehrlich-Schwoebel barrier in the system, we can apply the theory by Natori and Godby [69], which was presented in Sec. 1.4. This model can describe the rapid increase in anisotropy, but not explain the asymptotic behavior for higher vicinalities, even if the step-height transformation suggested by Pehlke and Tersoff [93] and measured by Tong and Bennett [145] is taken into account. The results lead in the correct direction and the inclusion of the step height transformation from single to double height steps improves the description of the results by Natori and Godby's theory [69]. As it is known, that also higher multi-steps can form [149], the exact (multi-)step structure has to be known to compare the aspect ratio of the anisotropic ICZs to the theory of Natori and Godby [69]. At this point, the question how the aspect ratio and the diffusion relate to each other is unclear and we tried to solve the question by numeric simulations. In this context, it was unclear, if the width of the island might at some point limit the width of the zone. If the width of the ICZ is limited by the island size, it might be more suitable to subtract both the island length in the ICZ long and short axis directions to receive relevant data. Our simulations show that both a subtracted island width and the pure ICZ aspect ratio show a square root dependence of the diffusion anisotropy, naturally with different pre-factors. We thus stayed with the numbers of the ICZs including the islands for consistency. We then used these simulations to rescale the data from the PEEM experiments for comparison with Natori and Godby's theory.

The inclusion of these simulation results then reduce the deviations from their theory and our experiments. We are certain that a single to multi-step height transformation and possibly facetting will have an influence on the theoretical predictions, especially as they are known to occur in the investigated vicinal Ag covered Systems close to the Si(0 0 1) orientation. Therefore, the facetting of the investigated vicinal surfaces was investigated using SPA-LEED at conditions as close as possible to those used for the PEEM experiments. The resulting step heights are used to recalculate the theory by Natori and Godby using the new mean terrace widths, assuming that the ESB is not largely influenced by the multi-step character of the step bundles. The PEEM experiments then fit almost perfectly to the theoretical predictions.

As the results can nearly perfectly explain the data for vicinal Si(0 0 1), we applied the theory to different substrate orientations as well. These are still on the same vicinality direction as the vicinal Si(0 0 1) samples. Again increase in diffusion anisotropy known from PEEM experiments for Ag/vic. Si(1 1 1) can be explained by the theory. The prediction is better again, if double height steps, as found in the literature, are assumed. These two sets of data both show an increasing diffusion anisotropy with decreasing difference in substrate orientation. In between these sets, some more low index surfaces have been investigated by PEEM, all showing the basically same isotropic diffusion that is only known for substrates with very low step density. The SPA-LEED experiments show, that here, facets in the [1 1 5] direction form and thus decrease the step density and increase the terrace width, even though they change the characteristics of the terraces. The only slightly deviating substrate orientation is the Si(0 0 1)-5° surface which already shows facets that are nevertheless not very well ordered. The diffusion anisotropy is lower than on the 4° vicinal Si(0 0 1) surface and the terrace size should be comparable to that on the 0.8° vicinal Si(0 0 1) surface. So far, the PEEM, SPA-LEED, and simulation data are in perfect agreement with the theoretical predictions and we can come to the conclusion, that we have reached a full understanding of the step induced diffusion anisotropy including facetting for Ag on vicinal Si(0 0 1) and Si(1 1 1) surfaces.

The next thing to be done is to check for the general applicability of this picture. To investigate the general applicability, the system of In on vicinal Si(0 0 1) was chosen as the decay mechanism was found to be similar to that of Ag/Si. Here, we are very surprised to measure isotropic ICZs for In/Si(0 0 1)-4° which for Ag was the surface with the highest diffusion anisotropy of all investigated surfaces. A similar situation was also found for several lower vicinalities. Nevertheless, this can again be explained by the same understanding we just gained from the Ag/Si systems. If the diffusion is isotropic for surfaces which naturally have a very high step density, three effects can cause the ICZ to be isotropic. First, the ICZ can be isotropic, because no ESB is present in the system. Second, the ICZ may be round as the temperature is too high and the ESB is already overcome at the investigated temperatures. Third, the step density is not as high as assumed, and for example reduced by facets or multi height steps. For the first and third case, we would expect that the ICZ remains round as we decrease the temperature in our desorption experiments, however, if an ESB is present and very low, we will most likely not see much of an effect when closing in on the lower end of the experimental temperature limits. This was in fact the case in the PEEM experiments. We did not see any change as a function of temperature and we can therefore not make any statement of the height of the ESB in this system. However, we were able to check for facetting. This time, the LEED images were recorded with µ-diffraction in the LEEM instrument. This showed, that after preparation, facets are present on the surface. These are also known in literature and run in the [310] direction. We thus can state, that also here, our understanding is in accordance to the experiments. As an additional result, we end up with some numbers for the diffusion of In/Si(001) as presented in Ch. 7.4.1.

The main result, however, is the answer to the question: How anisotropic is an anisotropic surface?

The answer is a clear: 'it depends'.

We can however now present the parameters on which it actually depends. First, the temperature has a strong influence on the diffusion anisotropy. Second, the terrace width and with it its diffusion anisotropy depends strongly on the investigated material system as facetting and multi-step formation can occur and also show a strong influence on the diffusion anisotropy.

Therefore, for the statement of how anisotropic an anisotropic surface is, knowledge about the step structure and ESB has to be obtained. If, however, both are known, a good prediction about the anisotropy of the diffusion anisotropy can be made. On the other hand, if one is only interested in the actual diffusion anisotropy, this can easily be confirmed by a quick and easy desorption experiment in PEEM or LEEM.

Our method to tackle surface diffusion is very well suited to study surface diffusion and diffusion anisotropy as it has access to a broad temperature range, is quick as the data are collected with video rate and has access to the surface structure in brightfield or darkfield LEEM modes. Furthermore, the facetting can be investigated by µ-diffraction and the accessible length scales allow statements about the entire surface.

9.3 Self Assembly of Single-Crystalline Ag Nanowires

It is known that nanowires form on $Si(0\,0\,1)$-4° in a self-organized manner. A discussion was found in literature about the origin of this nanowire formation. Tersoff and Tromp [124] suggested, that the wire formation is a strain release mechanism resulting in elongated islands. Others, such as Roos et al. [123] suggest, that the nanowire formation could also be triggered by diffusion anisotropy as this is very strong on the 4°vicinal $Si(0\,0\,1)$ surface. On this surface, all wires grow in the same direction, namely aligned with the steps and thus the fast diffusion direction, which might lead to preferred attachment at short side facets of the wires.

We also found wires on flat $Si(0\,0\,1)$ surfaces. Here, the wires are evenly distributed along the two 90° rotated dimer directions. On intermediate surface vicinalities, an increase in vicinality orders results in an ordering of the nanowire orientation. This transformation seems unaffected by the multi-step formation discussed previously (Ch. 6). The formation of these wires is temperature dependent with an activation energy of 1.4 eV, independent of the substrate's vicinality. Therefore, the diffusion anisotropy is an unlikely candidate to cause the formation of the wires. Furthermore, we observed shape transitions [158] similar to those previously predicted by simulations [124], which have been attributed to lattice mismatch between the Ag islands and the Si substrate. The measured activation energy for wire formation, however, is independent of the vicinality and stress is an unlikely candidate to cause the alignment of the wires with the steps as the strain should change with increasing vicinality.

Instead, the attachment rates of diffusing Ag atoms might very well influence the direction in which the wires elongate. Our other measurements (see Fig. 7.5) show, that the diffusion anisotropy on a $4°$ vicinal surface has an activation energy of $0.6\,eV$, consistent with the effects of an Ehrlich-Schwoebel barrier [70, 72]. Since the step density increases with the vicinality, the mass transport over large distances across the surface steps becomes increasingly hindered. On flat substrates, with well-spaced steps running in arbitrary directions [92, 55], the wires grow in one of the two possible directions with the same probability. But as the diffusion anisotropy and the step density grow and with an increased angle of vicinality, the influence of the Ehrlich-Schwoebel barrier becomes more dominant. We therefore conclude, that the nanowire's alignment is influenced by anisotropic diffusion which, for the case of $Ag/Si(0\,0\,1)$, is due to an ES-Barrier. The anisotropic material deposition along one of the substrate's principal directions must be sufficient to cause the wires to preferentially grow along the fast diffusion direction, namely parallel to the step edges, as the diffusion anisotropy increases.

Nevertheless, the measured increase of the ICZ's aspect ratio does not exhibit the linear dependence on the miscut that is found for the alignment of the nanowires (see Fig. 8.5). There is a clear indication, however, while the wire formation itself is most likely caused by lattice mismatch strain, that the ES-Barrier and the resulting diffusion anisotropy play a significant role in the predominant arrangement of the wires along the step edges for samples of higher vicinality. In total, we can not pick the two effects apart as strain causes anisotropic potentials and with it diffusion anisotropy and vice versa.

The ordering of the nanowires relates to the increase in diffusion anisotropy and we think this is the cause of wire orientation, yet there is no clear proof that this is also linked to the wire formation in the first place. For further investigations, the evolution of the aspect ratio could be investigated in the early growth stages as suggested by Tersoff and Tromp [124]. Such experiments have been carried out with a large field of view, nevertheless, reliable conclusions cannot be drawn from them as the resolution is not sufficient. For sufficient resolution, the field of view would have to be chosen to be rather small ($< 10\,\mu$m) and the probability that a wire nucleates within the field of view is very small. This thus is a tedious task to get statistically relevant data which did not succeed up to the present state. Nevertheless, these data will most likely not be proof that the suggested strain transformation is the only, but only one possible cause for nanowire growth in this system. We come to the conclusion, that further investigations on this matter are rather useless, as the two suggested effects to cause the formation of the nanowires are strongly linked to one another, and further investigations are a waste of good will.

Bibliography

[1] M. N. Baibich, J. M. Broto, A. Fert, F. N. Van Dau, F. Petroff, P. Etienne, G. Creuzet, A. Friederich, and J. Chazelas, Phys. Rev. Lett. **61**, 2472 (1988).

[2] G. Binasch, P. Grünberg, F. Saurenbach, and W. Zinn, Phys. Rev. B **39**, 4828 (1989).

[3] A. Fert, P. Grünberg, A. Barthélémy, F. Petroff, and W. Zinn, Journal of Magnetism and Magnetic Materials **140-144**, 1 (1995), international Conference on Magnetism.

[4] E. T. Swartz and R. O. Pohl, Rev. Mod. Phys. **61**, 605 (1989).

[5] P. E. Phelan, Journal of Heat Transfer **120**, 37 (1998).

[6] R. J. Stoner and H. J. Maris, Phys. Rev. B **48**, 16373 (1993).

[7] D. G. Cahill, W. K. Ford, K. E. Goodson, G. D. Mahan, A. Majumdar, H. J. Maris, R. Merlin, and S. R. Phillpot, Journal of Applied Physics **93**, 793 (2003).

[8] C. N. R. Rao, G. U. Kulkarni, P. J. Thomas, and P. P. Edwards, Chemistry: A European Journal **8**, 28 (2002).

[9] V. N. Lutskii, JETP Letters **2**, 245 (1965).

[10] A. B. Shick, J. B. Ketterson, D. L. Novikov, and A. J. Freeman, Phys. Rev. B **60**, 15484 (1999).

[11] G. Jnawali, Ph.D. thesis, Universität Duisburg-Essen, Fachbereich Physik (2009).

[12] S. Cahangirov, M. Topsakal, E. Aktürk, H. Şahin, and S. Ciraci, Phys. Rev. Lett. **102**, 236804 (2009).

[13] K. Novoselov, Nat Mater **6**, 720 (2007).

[14] Y. Zhang, T.-T. Tang, C. Girit, Z. Hao, M. C. Martin, A. Zettl, M. F. Crommie, Y. R. Shen, and F. Wang, Nature **459**, 820 (2009).

[15] H. A. Atwater, Scientific American Magazine **296**, 56 (2007).

[16] H. A. Atwater and A. Polman, Nat Mater **9**, 205 (2010).

[17] R. Zia, J. A. Schuller, A. Chandran, and M. L. Brongersma, Materials Today **9**, 20 (2006).

[18] S. A. Maier, Current Nanoscience **1**, 17 (2005).

[19] L. Chelaru and F. Meyer zu Heringdorf, Appl. Phys. Lett. **89**, 241908 (2006).

[20] L. Chelaru, M. Horn-von Hoegen, D. Thien, and F. Meyer zu Heringdorf, Phys. Rev. B **73**, 115416 (2006).

[21] L. Royer, Bull. Soc. Fr. Min. **51**, 7 (1928).

[22] T. Michely and J. Krug, *Islands, Mounds and Atoms Patterns and Processes in Crystal Growth Far from Equilibrium* (Springer-Verlag Berlin Heidelberg, 2004).

[23] E. Bauer, Zeitschrift für Kristallographie **110**, 372 (1958).

[24] W. F. Egelhoff and I. Jacob, Phys. Rev. Lett. **62**, 921 (1989).

[25] A. Einstein, Annalen der Physik **17**, 549 (1905).

[26] A. Fick, Annalen der Physik **94**, 59 (1855).

[27] J.-B.-J. Fourier, *Théorie analytique de la chaleur* (A Paris, Chez Firmiin Didot, Père et Fils, 1822).

[28] M. Bowker and D. A. King, Surface Science **71**, 583 (1978).

[29] A. G. Naumovets, Physica A: Statistical Mechanics and its Applications **357**, 189 (2005), diffusion and Soft Matter Physics - Proceedings of the 41st Karpacz Winter School of Theoretical Physics on Diffusion and Soft Matter Physics.

[30] G. A. Bassett, Philosophical Magazine **3**, 1042 (1958).

[31] H. Bethge, phys. stat. sol. b **2**, 775 (1962).

[32] J. Venables, G. Spiller, and M. Hanbuecken, Rep. Prog. Phys **47**, 399 (1984).

[33] J. A. Venables, Philosophical Magazine **27**, 697 (1973).

[34] J. Venables and G. L. Price, *Epitacial Growth*, vol. Part B (New York: Academic, 1975), ch. 4.

[35] B. Lewis and G. J. Rees, Philosophical Magazine **29**, 1253 (1974).

[36] B. Lewis and G. J. Rees, *Nucleation and Growth of Thin Films* (New York: Academic, 1978).

[37] G. Zinsmeister, Krist. Techn. **5**, 207 (1970).

[38] G. Zinsmeister, Thin Solid Films **2**, 497 (1968).

[39] G. Zinsmeister, Thin Solid Films **7**, 51 (1971).

[40] G. Zinsmeister, Vacuum **16**, 529 (1966).

[41] D. Walton, The Journal of Chemical Physics **37**, 2182 (1962).

[42] R. Vincent, Proceedings of the Royal Society of London. A. Mathematical and Physical Sciences **321**, 53 (1971).

[43] K. Routledge and M. Stowell, Thin Solid Films **6**, 407 (1970).

[44] M. Stowell and T. Hutchinson, Thin Solid Films **8**, 41 (1971).

[45] M. J. Stowell, Philosophical Magazine **21**, 125 (1970).

[46] E. W. Müller, R. Gomer, I. N. Stranski, and C. Herring, *The Use of Classical Macroscopic Concepts in Surface Energy Problems*, Structure and Properties of Solid Surfaces (University of Chicago Press, 1953).

[47] G. Wulff, Zeitschrift für Kristallographie **34**, 449 (1901).

[48] W. K. Burton, N. Cabrera, and F. C. Frank, Philosophical Transactions of the Royal Society of London. Series A, Mathematical and Physical Sciences **243**, 299 (1951).

[49] P. Nozières, *Shape and growth of Crystals*, Solids far from Equilibrium (Cambridge University Press, 1991).

[50] M. Giesen, Progress in Surface Science **68**, 1 (2001).

[51] C. Herring, *The use of Classical Macroscopic Concepts in Surface Energy Problems*, Structure Properties of Solid Surfaces (University of Chicago Press, 1953).

[52] H.-C. Jeong and E. D. Williams, Surface Science Reports **34**, 171 (1999).

[53] S. T. Chui and J. D. Weeks, Phys. Rev. B **14**, 4978 (1976).

[54] J. Lapujoulade, Surface Science Reports **20**, 195 (1994).

[55] R. M. Tromp and M. Reuter, Phys. Rev. Lett. **68**, 820 (1992).

[56] J. E. Avron, H. van Beijeren, L. S. Schulman, and R. K. P. Zia, Journal of Physics A: Mathematical and General **15**, L81 (1982).

[57] E. Gruber and W. Mullins, Journal of Physics and Chemistry of Solids **28**, 875 (1967).

[58] C. Jayaprakash, C. Rottman, and W. F. Saam, Phys. Rev. B **30**, 6549 (1984).

[59] M. P. A. Fisher, D. S. Fisher, and J. D. Weeks, Phys. Rev. Lett. **48**, 368 (1982).

[60] M. E. Fisher and D. S. Fisher, Phys. Rev. B **25**, 3192 (1982).

[61] Y. Akutsu, N. Akutsu, and T. Yamamoto, Phys. Rev. Lett. **61**, 424 (1988).

[62] T. Yamamoto, Y. Akutsu, and N. Akutsu, Journal of the Physical Society of Japan **63**, 915 (1994).

[63] E. D. Williams, R. Phaneuf, J. Wei, N. Bartelt, and T. Einstein, Surface Science **310**, 451 (1994).

[64] A. F. Andreev, Sov. Phys. JETP **53**, 1063 (1981).

[65] C. Herring, Phys. Rev. **82**, 87 (1951).

[66] J. M. Blakely, *Introduction to the properties of crystal surfaces*, vol. 1st edition (Pergamon Press, 1973).

[67] J. Cahn, *Interfacial Segregation* (American Society of Metals, 1977).

[68] E. D. Williams and N. C. Bartelt, *Handbook of Surface Science* (North-Holland, Amsterdam, 1996).

[69] A. Natori and R. W. Godby, Phys. Rev. B **47**, 15816 (1993).

[70] R. L. Schwoebel and E. J. Shipsey, Journal of Applied Physics **37**, 3682 (1966).

[71] R. L. Schwoebel, Journal of Applied Physics **40**, 614 (1969).

[72] G. Ehrlich and F. Hudda, The Journal of Chemical Physics **44**, 1039 (1966).

[73] U. Scheithauer, G. Meyer, and M. Henzler, Surface Science **178**, 441 (1986).

[74] M. Horn-von Hoegen, Zeitschrift fuer Kristallographie **214**, 1 (1999).

[75] Omicron Nanotechnology GmbH, *ELECTRON GUN Instructions Manual*.

[76] Omicron Nanotechnology GmbH, *SPECTALEED OPTICS Manual*.

[77] M. P. Seah and W. A. Dench, Surface and Interface Analysis **1**, 2 (1979).

[78] F. Meyer zu Heringdorf and M. Horn-von Hoegen, Review of Scientific Instruments **76**, 08512 (2005).

[79] P. Kury, Ph.D. thesis, Universität Duisburg-Essen (2007).

[80] R. Tromp and M. Reuter, Ultramicroscopy **36**, 99 (1991).

[81] T. Duden, Vakuum in Forschung und Praxis **17**, 92 (2005).

[82] Supplier's Website, URL `http://www.crystec.de/silizium-d.html`.

[83] Merfen, Burkhard (1999), hausarbeit im Rahmen der Ersten Staatsprüfung für das Lehramt an Gymnasien, Institut für Festkörperphysik Universität Hannover.

[84] J. C. Sturm and C. M. Reaves, Electron Devices, IEEE Transactions on **39**, 81 (1992).

[85] URL `www.webelements.com`.

[86] G. Chiarotti, *Landolt Börnstein- Group III Condensed Matter: Numerical Data and Functional Relationships in Science and Technology*, vol. 24a: Structure (Springer, 2011).

[87] C. R. Hubbard, H. E. Swanson, and F. A. Mauer, Journal of Applied Crystallography **8**, 45 (1975).

[88] S. M. Sze and K. N. Kwok, *Physics of Semiconductor Devices* (John Wiley and Sons, Inc., 1981), 3rd ed.

[89] K. Takayanagi, Y. Tanishiro, S. Takahashi, and M. Takahashi, Surf. Sci. **164**, 367 (1985).

[90] R. E. Schlier and H. E. Farnsworth, The Journal of Chemical Physics **30**, 917 (1959).

[91] R. M. Tromp, R. J. Hamers, and J. E. Demuth, Phys. Rev. Lett. **55**, 1303 (1985).

[92] J. Tersoff and E. Pehlke, Phys. Rev. Lett. **68**, 816 (1992).

[93] E. Pehlke and J. Tersoff, Phys. Rev. Lett **67**, 465 (1991).

[94] W. Ranke, Phys. Rev. B **41**, 5243 (1990).

[95] F. Meyer zu Heringdorf, N. Buckanie, L. Chelaru, and N. Raß, *EMC 2008 14th European Microscopy Congress 1-5 September 2008, Aachen, Germany* (Springer Berlin Heidelberg, 2008), chap. Imaging of Surface Plasmon Waves in Nonlinear Photoemission Microscopy, p. 737.

[96] H. Ehrenreich and H. R. Philipp, Phys. Rev. **128**, 1622 (1962).

[97] K. N. Tu, Journal of Applied Physics **94**, 5451 (2003).

[98] Z. Suo and W. Wang, Journal of Applied Physics **76**, 3410 (1994).

[99] E. Arzt, O. Kraft, W. D. Nix, and J. J. E. Sanchez, Journal of Applied Physics **76**, 1563 (1994).

[100] T. Marieb, P. Flinn, J. C. Bravman, D. Gardner, and M. Madden, Journal of Applied Physics **78**, 1026 (1995).

[101] B. Stahlmecke, F.-J. Meyer zu Heringdorf, L. I. Chelaru, M. Horn-von Hoegen, G. Dumpich, and K. R. Roos, Appl. Phys. Lett. **88**, 053122 (2006).

[102] A. Saranin, A. Zotov, V. Lifshits, M. Katayama, and K. Oura, Surf. Sci. **426**, 298 (1999).

[103] L. Lottermoser, E. Landemark, D. Smilgies, M. Nielsen, R. Feidenhans'l, G. Falkenberg, R. Johnson, M. Gierer, A. Seitsonen, H. Kleine, et al., Phys. Rev. Lett. **80**, 3980 (1998).

[104] P. Eckerlin and H. Kandler, *Landolt Börnstein- Group III Condensed Matter: Numerical Data and Functional Relationships in Science and Technology*, vol. 6: Structure Data of Elements and Intermetallic Phases (Springer, 2011).

[105] L. Liu and W. Bassett, Journal of Applied Physics **44**, 1475 (1973).

[106] M. Weinert and R. E. Watson, Phys. Rev. B **29**, 3001 (1984).

[107] R. S. Seth and S. B. Woods, Phys. Rev. B **2**, 2961 (1970).

[108] R. O. Simmons and R. W. Balluffi, Phys. Rev. **119**, 600 (1960).

[109] K. Wan, X. Lin, and J. Nogami, Physical Review B **47**, 13700 (1993).

[110] Y. Ding, C. Chan, and K. Ho, Phys. Rev. Lett. **67**, 1454 (1991).

[111] G. Le Lay, Surf. Sci. **132**, 169 (1983).

[112] R. M. Tromp, Y. Fujikawa, J. B. Hannon, A. W. Ellis, A. Berghaus, and O. Schaff, Journal of Physics: Condensed Matter **21**, 314007 (2009).

[113] D. Spoddig, C. Hassel, and M. Farle, eBSD measurements.

[114] G. Friedel, *Lecons de Cristallographie* (1925), paris.

[115] M. Kronberg and M. Wilson, Trans. Am. Inst. Min. (Metall) Engrs. **185**, 501 (1949).

[116] D. H. Warrington, Journal de Physique **C4**, 87 (1975).

[117] M. Baake, arxiv:math.MG/0605222 v1.

[118] R. V. Moody, ed., *The Mathematics of Long-Range Aperiodic Order*, NATO Asi Series, NATO (Springer Netherlands, 1997).

[119] Y. Ikuhara and P. Pirouz, Microscopy Research and Technique **40**, 206 (1998).

[120] Y. Ikuhara and P. Pirouz, Materials Science Forum **207**, 121 (1996).

[121] H. Fecht, Philosophical Magazine Letters **60**, 81 (1989).

[122] I. Amidror, *The Theory of the Moiré Phenomenon Volume I: Periodic Layers*, vol. 38 of *Computational Imaging and Vision* (Springer London, 2009), 2nd ed.

[123] K. R. Roos, K. L. Roos, M. Horn-von Hoegen, and F.-J. Meyer zu Heringdorf, J. Phys.: Condens. Matter **17**, S1407 (2005).

[124] J. Tersoff and R. Tromp, Phys. Rev. Lett. **70**, 2782 (1993).

[125] T. Fukuda, Phys Rev. B **50**, 1969 (1994).

[126] C. Collazo-Davila, D. Grozea, and L. Marks, Phys. Rev. Lett. **80**, 1678 (1998).

[127] W. Fan and A. Ignatiev, Phys. Rev. B **41**, 3592 (1990).

[128] G. Le Lay, M. Manneville, and R. Kern, Surface Science **72**, 405 (1978).

[129] D. Wall, K. R. Roos, M. Horn-von Hoegen, and F.-J. Meyer zu Heringdorf, Mater. Res. Soc. Symp. Proc. **1088**, 1088 (2008).

[130] I. Lohmar, Ph.D. thesis, University of Cologne (2009).

[131] K. R. Roos, K. L. Roos, I. Lohmar, D. Wall, J. Krug, M. Horn-von Hoegen, and F.-J. Meyer zu Heringdorf, Physical Review Letters **100**, 016103 (2008).

[132] N. Ferralis, F. E. Gabaly, A. K. Schmid, R. Maboudian, and C. Carraro, Phys. Rev. Lett. **103**, 256102 (2009).

[133] G. Raynerd, M. Hardiman, and J. Venables, Phys. Rev. B **44**, 13803 (1991).

[134] M. Hanbücken, T. Doust, and J. A. V. O. Osasona, G. Le Lay, Surf. Sci. **168**, 133 (1986).

[135] H. Jeong and S. Jeong, Phys. Rev. B **71**, 035310 (2005).

[136] J. Mysliveček, P. Sobotík, I. Ošt'ádal, T. Jarolímek, and P. Šmilauer, Phys. Rev. B **63**, 045403 (2001).

[137] P. Sobotík, P. Kocán, and I. Ost'ádal, Surface Science **537**, L442 (2003).

[138] H. Yeom, I. Matsuda, K. Tono, and T. Ohta, Phys. Rev. B **57**, 3949 (1998).

[139] X. Lin, K. Wan, and J. Nogami, Phys. Rev. B **49**, 7385 (1994).

[140] T. Michely, M. Reuter, M. Copel, and R. Tromp, Phys. Rev. Lett. **73**, 2095 (1994).

[141] D. Wall, Diploma-Thesis, Universität Duisburg-Essen (2007).

[142] K. Kong, H. Yeom, D. Ahn, H. Yi, and B. Yu, Physical Review B (Condensed Matter and Materials Physics) **67**, 235328 (2003).

[143] S. Fölsch, G. Meyer, K. H. Rieder, M. Horn-von Hoegen, T. Schmidt, and M. Henzler, Surface Science **394**, 60 (1997).

[144] S. Fölsch, D. Winau, G. Meyer, K. Rieder, M. Horn-von Hoegen, T. Schmidt, and M. Henzler, Applied Physics Letters **67**, 2185 (1995).

[145] X. Tong and P. A. Bennett, Phys. Rev. Lett. **67**, 101 (1991).

[146] F. Meyer zu Heringdorf, T. Schmidt, S. Heun, R. Hild, P. Zahl, B. Ressel, E. Bauer, and M. Horn-von Hoegen, Phys. Rev. Lett. **86**, 5088 (2001).

[147] O. L. Alerhand, A. N. Berker, J. D. Joannopoulos, D. Vanderbilt, R. J. Hamers, and J. E. Demuth, Phys. Rev. Lett **64**, 2406 (1990).

[148] O. L. Alerhand, A. Nihat Berker, J. D. Joannopoulos, D. Vanderbilt, R. J. Hamers, and J. E. Demuth, Phys. rev. Lett. **66**, 962 (1991).

[149] A. Meier, P. Zahl, R. Vockenroth, and M. Horn-von Hoegen, Appl. Surf. Sci. **123/124**, 694 (1998).

[150] M.-T. Hütt, *Datenanalyse in der Biologie* (Springer Verlag Berlin, 2001).

[151] C. Klein, T. Nabbefeld, H. Hattab, D. Meyer, G. Jnawali, M. Kammler, F.-J. M. zu Heringdorf, A. Golla-Franz, B. H. Müller, T. Schmidt, et al., Review of Scientific Instruments **82**, 035111 (2011).

[152] D. Chadi, Phys. Rev. Lett. **59**, 1691 (1987).

[153] F.-J. Meyer zu Heringdorf, Ph.D. thesis, Universität Hannover (1999).

[154] J. Knall, S. Barnett, J.-E. Sundgren, and J. Greene, Surface Science **209**, 314 (1989).

[155] R. Phaneuf, E. Williams, and N. Bartelt, Physical Review B **38**, 1984 (1988).

[156] R. J. Phaneuf and E. D. Williams, Phys. Rev. B **41**, 2991 (1990).

[157] R. Phaneuf and E. Williams, Phys. Rev. Lett. **58**, 2563 (1987).

[158] S. Sindermann, D. Wall, K. R. Roos, M. Horn-von Hoegen, and F.-J. Meyer zu Heringdorf, Electronic Journal of Surface Science and Nanotechnology **8**, 372 (2010).

[159] B. Li, W. Swiech, and J. Zuo, Microsc Microanal **13**, 726 (2007).

[160] W. Cho, J. Kim, N. Park, K. Chae, J. Kim, I. Lyo, S. Kim, D. Choi, and C. Whang, Surface Science **439**, L792 (1999).

[161] Y. W. Kim, N. G. Park, W. S. Cho, K. H. Chae, C. N. Whang, K. S. Kim, S. S. Kim, and D. S. Choi, Surf. Sci. **396**, 295 (1998).

[162] E. Bauer, Surface Science **299/300**, 102 (1994).

Appendix A

Publications

A.1 Peer-Reviewed Articles

1. J. Coraux, A. T. N'Diaye, M. Engler, C. Busse, D. Wall,
   N. M. Buckanie, F.-J. Meyer zu Heringdorf, R. van Gastel,
   B. Poelsema, and T. Michely, New Journal of Physics **11**,
   023006 (2009).

2. J. Coraux, A. T. N'Diaye, M. Engler, C. Busse, D. Wall,
   N. M. Buckanie, F.-J. Meyer zu Heringdorf, R. van Gastel,
   B. Poelsema, and T. Michely, New Journal of Physics **11**,
   039801 (2009).

3. H. Hattab, A. T. N'Diaye, D. Wall, G. Jnawali, J. Coraux,
   C. Busse, R. van Gastel, B. Poelsema, T. Michely, F.-J.

Meyer zu Heringdorf, and M. Horn-von Hoegen, Applied Physics Letters **98**, 141903(2011).

4. G. Jnawali, F.-J. Meyer zu Heringdorf, D. Wall, S. Sindermann, and M. Horn von Hoegen, J. Vac. Sci Technol. B **27**, 180 (2009).

5. P. Kury, K. R. Roos, D. Thien, S. Möllenbeck, D. Wall, M. Horn-von Hoegen, and F.-J. Meyer zu Heringdorf, Organic Electronics **9**, 461 (2008).

6. A. T. N'Diaye, R. van Gastel, A. J. Martinez-Galera, J. Coraux, H. Hattab, D. Wall, F.-J. Meyer zu Heringdorf, M. Horn-von Hoegen, J. M. Gomez-Rodriguez, B. Poelsema, C. Busse, and T. Michely, New Journal of Physics **11**, 113056 (2009).

7. K. R. Roos, K. L. Roos, I. Lohmar, D. Wall, J. Krug, M. Horn-von Hoegen, and F.-J. Meyer zu Heringdorf, Physical Review Letters **100**, 016103 (2008).

8. S. Sindermann, D. Wall, K. R. Roos, M. Horn-von Hoegen, and F.-J. Meyer zu Heringdorf, Electronic Journal of Surface Science and Nanotechnology **8**, 372 (2010).

9. R. van Gastel, A. T. N'Diaye, D. Wall, J. Coraux, C. Busse, N. M. Buckanie, F.-J. Meyer zu Heringdorf, M. Horn von Hoegen, T. Michely, and B. Poelsema, Appl. Phys. Lett. **95**, 121901 (2009).

10. D. Wall, S. Tikhonov, S. Sindermann, D. Spoddig, C. Hassel, M. Horn- von Hoegen, and F.-J. Meyer zu Heringdorf, IBM Journal of Research and Development **55**(4), 9:1 (2011)

11. D. Wall, S. Sindermann, K. R. Roos, M. Horn-von Hoegen, and F.-J. Meyer zu Heringdorf, J. Phys.: Condens. Matter **21**, 314023 (2009).

12. D. Wall, K. R. Roos, M. Horn-von Hoegen, and F.-J. Meyer zu Heringdorf, Mater. Res. Soc. Symp. Proc. **1088**, 1088-W05-04 (2008).

13. D. Wall, I. Lohmar, K. R. Roos, J. Krug, M. Horn-von Hoegen, and F.-J. Meyer zu Heringdorf, New Journal of Physics **12**, 103019 (2010)

A.2 Conference Contributions

A.2.1 Talks

1. F.-J. Meyer zu Heringdorf, D. Wall, S. Sindermann, M. Horn-von Hoegen, I. Lohmar, J. Krug, and Roos K. R., The Balance of Diffusion and Desorption around Islands on Si: Diffusion Made Visible, DPG Frühjahrstagung Sektion Kondensierte Materie Post Deadline Session (2010).

2. S. Möllenbeck, D. Thien, Kury. P., K. R. Roos, D. Wall, M. Horn-von Hoegen, and F.-J. Meyer zu Heringdorf, A PEEM Study of the Substrate Dependence of Pentacene

Thin Film Growth on Silicon, DPG Frühjahrstagung Sektion Kondensierte Materie O63.3 (2008).

3. A. T. N'Diaye, J. Coraux, R. Djemour, D. Wall, N. M. Buckanie, F.- J. Meyer zu Heringdorf, R. van Gastel, B. Poelsema, T. Michely, and C. Busse, Compressive strain relaxation through wrinkle formation in epitaxial graphene on Ir(111), DPG Frühjahrstagung Sektion Kondensierte Materie O34.3 (2009).

4. S. Sindermann, D. Wall, M. Horn-von Hoegen, and F.-J. Meyer zu Heringdorf, High temperature growth of Ag-nanowires on bare Si(001), DPG Frühjahrstagung Sektion Kondensierte Materie O5.2 (2009).

5. D. Wall, I. Lohmar, K. R. Roos, J. Krug, M. Horn-von Hoegen, and F.-J. Meyer zu Heringdorf, High temperature Surface Diffusion Involving Multiple Reconstructions: Ag/Si(111), 17th International Microscopy Congress I11.2 (2010).

6. D. Wall, I. Lohmar, K. R. Roos, J. Krug, M. Horn-von Hoegen, and F.-J. Meyer zu Heringdorf, Real Time Measurements of Surface Diffusion in a Multi- Reconstruction System: Ag/Si(111), DPG Frühjahrstagung Sektion Kondensierte Materie O68.2 (2010).

7. D. Wall, K. R. Roos, K. L. Roos, I. Lohmar, J. Krug, M. Horn-von Hoegen, and F.-J. Meyer zu Heringdorf, A Real

Time Investigation of Surface Diffusion Fields in Epitaxially Grown Systems, Materials Research Society Spring Meeting: Symposium W (2008).

8. D. Wall, K. R. Roos, K. L. Roos, I. Lohmar, J. Krug, M. Horn-von Hoegen, and F.-J. Meyer zu Heringdorf, Real Time Imaging of Surface Diffusion Fields during Island decay, DPG Frühjahrstagung Sektion Kondensierte Materie O33.3 (2008).

9. D. Wall, K. R. Roos, K. L. Roos, I. Lohmar, J. Krug, M. Horn-von Hoegen, and F.-J. Meyer zu Heringdorf,, A Real Time Investigation of Surface Diffusion Fields in Epitaxially Grown Systems, European Conference on Surface Science 26 WeA-A- DIF3-266 (2009).

10. Dirk Wall, Graphene Prospects for Time-Resolved PEEM, 2nd NationalWorkshop on Non-Linear and Time-Resolved PEEM (2009).

11. Dirk Wall, Growth and Decay of Silver Islands, Deutsche Gesellschaft für Kristallwachstum und Kristallzüchtung: 9. Kinetik-Seminar (2008), (invited).

12. D. Wall, S. Sindermann, K. R. Roos, I. Lohmar, J. Krug, M. Horn-von Hoegen, F.-J. Meyer zu Heringdorf, Dynamics of Reconstructed Zones Formed Around Islands on Si During Desorption: Diffusion Made Visible, 8th International Symposium on Atomic Level Characterizations for New Materials and Devices '11 26SS07 (2011).

A.2.2 Poster

1. D. Wall, S. Sindermann, M. Horn-von Hoegen, and F.-J. Meyer zu Heringdorf, The interplay of Anisotropic Diffusion Fields and Ag Nanowire formation on vicinal Si(001), LEEM PEEM 6 Conference (2008).

2. D. Wall, S. Sindermann, M. Horn-von Hoegen, and F.-J. Meyer zu Heringdorf, The influence of diffusion anisotropy and strain on Ag nanowire formation on Flat and vicinal Si(001), DPG Frühjahrstagung Sektion Kondensierte Materie O27.54 (2009).

3. D. Wall, et. al., various Posters at National and International Workshops of SFB616: Energy dissipation at Surfaces (2007-2010).

A.2.3 Non-Peer-Reviewed Articles

1. A. T. N'Diaye, R. van Gastel, A. J. Martinez-Galera, J. Coraux, H. Hattab, D. Wall, F.-J. Meyer zu Heringdorf, M. Horn-von Hoegen, J. M. Gomez- Rodriguez, B. Poelsema, C. Busse, and T. Michely, Proceedings of the International Microscopy Congress 17 (IMC-17) (2010).

2. D. Wall, I. Lohmar, K. R. Roos, J. Krug, M. Horn-von Hoegen, and F.-J. Meyer zu Heringdorf, Proceedings of the International Microscopy Congress 17 (IMC-17) (2010).

3. F.-J. Meyer zu Heringdorf, D. Wall, K. R. Roos, I. Lohmar, J. Krug and M. Horn-von Hoegen, Proceedings of the 8th

International Symposium on Atomic Level Characterizations for New Materials and Devices '11 (2011)

Aknowledgements

At the end of this work, I would like to use the chance to acknowledge everyone who has directly or indirectly influenced and supported me in the course of this work.
I would like to thank Prof. Dr. M. Horn-von Hoegen for giving me the chance to work on this thesis in his group.
Many thanks have to go to Dr. F.-J. Meyer zu Heringdorf for the experimental and scientific support in the course of this work. He has also supported me with many good ideas, a friendly atmosphere and throughout many discussions.
I also have to thank K.R. Roos, I. Lohmar and J. Krug for many fruitful discussions on our way to many succesful publications.
I also want to thank the people who read through this work on the hunt for errors and typos: Claudius, Kelly, Simone.
Furthermore, I have to thank the two Diploma-Students I had the honor to guide throughout their thesis. S. Sindermann was very thorough and patient when counting thousands and thousands of islands and wires from the PEEM and LEEM experi-

ments. S. Thikonov has also done a lot of experimental work on the SPA-LEED instrument.

I also have to thank L. Kujawinski and H. Wolf for the technical support whenever needed and much fun throughout the work on this thesis.

I want to thank my parents for their support throughout my life, which from the beginning layed the path to this thesis.

Of course do I need to thank my wife Simone for the scientific and private support since the beginning of my academic studies.